CAMBRIDGE LIBRARY COLLECTION

Books of enduring scholarly value

Darwin

Two hundred years after his birth and 150 years after the publication of 'On the Origin of Species', Charles Darwin and his theories are still the focus of worldwide attention. This series offers not only works by Darwin, but also the writings of his mentors in Cambridge and elsewhere, and a survey of the impassioned scientific, philosophical and theological debates sparked by his 'dangerous idea'.

A Syllabus of a Course of Lectures on Mineralogy

John Stevens Henslow (1796–1861) was introduced to geology by Adam Sedgwick in 1819, and became Professor of Mineralogy at Cambridge in 1822. He soon moved to a chair in Botany, and not long afterwards became a teacher and mentor to Charles Darwin. This book on mineralogy was first published in 1823. It was intended to save time in class by providing an easily accessible reference to the composition of various minerals according to the principles of atomic theory, which was then entering the scientific mainstream. In that paradigm, analysis and examination of any mineral's composition involved first ascertaining the mineral's elementary molecules, both 'essential' and 'accidental', and second, determining the proportions in which the essential ingredients united to form the integrant molecule of the mineral. Henslow's book will interest historians of science tracing the development of atomic theory, and those working more broadly in the history of university education and the intellectual climate of the nineteenth century.

Cambridge University Press has long been a pioneer in the reissuing of out-of-print titles from its own backlist, producing digital reprints of books that are still sought after by scholars and students but could not be reprinted economically using traditional technology. The Cambridge Library Collection extends this activity to a wider range of books which are still of importance to researchers and professionals, either for the source material they contain, or as landmarks in the history of their academic discipline.

Drawing from the world-renowned collections in the Cambridge University Library, and guided by the advice of experts in each subject area, Cambridge University Press is using state-of-the-art scanning machines in its own Printing House to capture the content of each book selected for inclusion. The files are processed to give a consistently clear, crisp image, and the books finished to the high quality standard for which the Press is recognised around the world. The latest print-on-demand technology ensures that the books will remain available indefinitely, and that orders for single or multiple copies can quickly be supplied.

The Cambridge Library Collection will bring back to life books of enduring scholarly value across a wide range of disciplines in the humanities and social sciences and in science and technology.

A Syllabus of a
Course of Lectures
on Mineralogy

JOHN STEVENS HENSLOW

CAMBRIDGE
UNIVERSITY PRESS

CAMBRIDGE UNIVERSITY PRESS

Cambridge New York Melbourne Madrid Cape Town Singapore São Paolo Delhi

Published in the United States of America by Cambridge University Press, New York

www.cambridge.org
Information on this title: www.cambridge.org/9781108002011

This edition first published 1823
This digitally printed version 2009

ISBN 978-1-108-00201-1

SYLLABUS.

A

SYLLABUS

OF

𝔄 Course of Lectures

ON

MINERALOGY.

———◆◆———

BY

JOHN STEVENS HENSLOW,

M.A. M.G.S. & F.L.S.

PROFESSOR OF MINERALOGY IN THE UNIVERSITY OF CAMBRIDGE;

SECRETARY TO THE CAMBRIDGE PHILOSOPHICAL SOCIETY.

————————

Cambridge:

Printed by J. Hodson, Trinity-Street,

AND SOLD BY J. DEIGHTON AND SONS.

————

1823.

PREFACE.

SOME explanation seems necessary for having deviated from the plan pursued by my predecessor, in forming his Syllabus of a Course of Lectures for the University.

Having compared the analyses (given by different authors) of various minerals, with the views established by the Atomic Theory, the results of some of these investigations appeared sufficiently important to merit the attention of those who might wish to pursue this subject; but as they are confined to calculations, which it would have been tedious both for a Lecturer to dictate and for his audience to transcribe, it seemed preferable to print them. From a similar consideration, the specific gravities and other properties, which require numerical exposition, have been inserted in the Syllabus. Since, however, these calculations, if added to the notices of the different subjects to be detailed under each species, would have augmented the size of a work of this nature considerably beyond its ordinary limits, it became necessary to deviate entirely from the mode originally adopted by Dr. CLARKE. I have therefore supposed the description of each species to be divided into nine different heads, and the objects which each head is intended to include are collected between pages 2 and 6. From these will be selected, for every species, whatever may appear most worthy of engaging the attention; although reference is seldom made in the remainder of the Syllabus to any but such as embrace the description of some peculiar form, or involve numerical explanation.

The figures in page 3 are intended as general representations of a rhomboid, quadrangular prism, and octohedron; to these, for the sake of brevity in denoting the angles of the primitive forms, reference is occasionally made throughout the Syllabus, (under head II. for each species.) The mention of the regular solids of course does not stand in need of such assistance.

Many of the theoretic results, as well as some other notices, are not to be considered as conclusions upon which any reliance can be fixed, or as exhibiting my own opinions upon the subject. They are to be viewed merely as memoranda, printed for the sake of saving trouble in the lecture-room. Neither are the arrangement and nomenclature intended to form a sketch of any system of Mineralogy. The order pursued is that which appeared best adapted, in a short and elementary course of lectures for the University, to impress some knowledge of minerals upon those who may wish to acquire it, and assist others in their desire of prosecuting the study.

Perhaps it will not be considered misplaced if I here mention the methods which have been adopted in making the calculations alluded to, since they appear to be the most convenient for investigating the composition of any mineral according to the principles of the Atomic Theory. In the examination of any analysis there are two questions to be solved. First, to determine the proximate principles, or state of chemical combination in which the elementary particles exist in the mineral; in other words, to ascertain its

component, (or, as they are usually styled, elementary) molecules, both essential and accidental.
Secondly, to select the essential ingredients, and
ascertain the proportions in which they unite to
form the integrant molecule. Without considering
the various collateral circumstances which might
influence the determination of the first question,
we will speculate chiefly upon that evidence which
may be drawn from the proportions between the
different ingredients obtained by analysis. Taking
the Bournonite as an example, we have,

	Analysis.	Rep. Nos.	Proportion of Atoms.		
Antimony	24.23	5.625	4.30 +	430	
Lead	42.62	13.	3.27 +	327	
Copper	12.8	8.	1.60	160	nearly
Iron	1.2	3.5	.34 +	34	
Sulphur	17.	2.	8.50	850	

$$\left. \begin{array}{l} 4.30+ \\ 3.27+ \\ 1.60 \\ .34+ \\ 8.50 \end{array} \right\} = \left. \begin{array}{l} 430 \\ 327 \\ 160 \\ 34 \\ 850 \end{array} \right\} \text{nearly}$$

The first column of this Table contains the
analysis given by Hatchet in the Philosophical
Transactions (1804), quoted in page 41 of the
Syllabus. In the second column are arranged
the representative numbers (or weights of the
atoms) of each substance, selected from the
Table given at the end of this Preface.—
Dividing the weights in the first column by the
respective numbers in the second, the results

(placed in the third) shew the proportion between the *whole number of atoms* of the different substances contained in the mineral. And here, we may observe, that, when we have ascertained the component molecules, it is to this column we should look for the ingredient which is to mark the genus of the mineral, rather than to the weights in the original analysis. For we may fairly conclude that the character of the mineral will, *cæteris paribus*, be stamped by that substance which employs the greatest number of atoms towards its composition. Where the contrary is the case, of which an example will presently be given, we must look to other causes for explaining the phenomena. Should the representative number of that substance which employs the greater number of atoms be considerably less than the representative number of one of the other substances, it may happen that the weight of the former ingredient will be much less than that of the latter; which is the case in the example selected. Omitting the sulphur, which cannot mark the genus, we have among the four metals, the atoms of Antimony to those of Lead nearly as 430 : 327, whilst the weights of those metals are as 242 : 426 (nearly).   Hence I would rather class this substance with the ores of Antimony than with those of Lead, with which it has been usual to place it.

It appears that the sum of the atoms of the four metals does not greatly exceed those of the sulphur; we may therefore conjecture that each metal exists in the form of a sulphuret. Calculating the quantity of sulphur requisite to fulfil this condition, the result given in page 41 is obtained, by which it appears that the sulphur required (theoretically) exceeds the quantity actually obtained by experiment by $2.04 +$; but as this error should not fall upon the sulphur alone, the mean error for five ingredients would be only $.40 +$.*

In calculating the quantity of sulphur requisite, we may multiply the atoms of the respective ingredients by 2 (the representative number for sulphur); but, in general, the more correct method is to obtain it by a simple proportion, which avoids

---

* This exposition is not strictly accurate, since the error, if divided among five ingredients of different representative numbers and weights, would be greater for some than others; but it is sufficiently near to strengthen (as a general conclusion) the probability that the mineral is composed of four *sulphurets*.

the possibility of error arising from a neglect of
decimals beyond the tenth or hundredth places.

These calculations are founded upon the supposition that the original analysis is perfectly
correct, but it is evident that one source of error
may arise from the adoption of different representative numbers from those used in calculating
for the analysis from the various precipitates
obtained by experiment. For example, if the
original calculations in this analysis of the Bournonite had been made with the numbers here
adopted, it would have stood,

Analysis.		Precipitates.
Antimony 23.23 derived from		Per-ox. Ant.... 31.5
Lead 39.06		Sulphate Lead 60.1
Copper.... 12.8		Per-ox.Copper 16.
Iron 1.2		Iron.............. 1.2
Sulphur .. 17.		Sulphur .... 17.

and consequently the proportion between the
atoms of the whole, would have been (nearly) as,

Antimony............412 ⎫
Lead ....300 ⎪
Copper................160 ⎬
Iron .... 34 ⎪
Sulphur ....850 ⎭

which will not affect the conclusion originally
established, that each metal is united with the

sulphur, whilst it actually brings the sum of their atoms still nearer to those of the sulphur. As it would have been impossible to have reduced many of the original analyses in this manner, without more trouble than the subject merits, these are, for the most part, presumed to be strictly accurate, and the error will seldom be found of sufficient magnitude to affect the general result, unless where late experiments have led to a considerable change in the value of a representative number.

The second part of the problem we are attempting to solve, is to estimate the ingredients (and the proportion of each) *essential* to the composition of the integrant molecule which stamps the crystalline character of the mineral. The views of Haüy on this subject have been adopted, which ascribe to every different form in the integrant molecule a peculiar combination of elementary bodies. The doctrine of Isomorphic bodies, supported by Berzelius, and strenuously opposed in the last edition of Haüy's Mineralogy, seems as yet to rest upon such shallow foundation, that it would be premature to consider it as established, until it shall have received a greater accession of strength, or until it shall be shewn that the more simple hypothesis is untenable. Nothing can be easier than to adapt a set of the symbols invented by Berzelius to any analysis, and if we are to suppose that all (or even, in some cases, the greater part) of the ingredients exhibited by every accurate analysis of the same mineral to be essential to the composition of the integrant molecule, crystallographically considered, we must no longer expect precision in this department of Natural History. In some cases, the evidence for the adventitious character of an ingredient is clear; in Fontainbleau Spar (see page 63), the weight of the silica forms more than two-thirds of the mass, and the atoms of silica are to those of the carbonate of lime in a greater ratio than 6 : 1, yet no doubts are entertained of the crystalline character having been impressed by the smaller ingredient. The difficulty of conceiving the form of certain homogeneous and transparent minerals to have been impressed by a small portion of their ingredients, may perhaps be diminished by considering that various substances are capable of holding others in *solution,* where we suppose an admixture perfectly homogeneous to take place without the formation of any new integrant molecule, and it does not seem improbable that this principle may admit of further extension than has hitherto been ascribed to it. In this light, steel might be considered as a solution of carbon in iron; plumbago a solution of iron in carbon; naptha a solution of carbon in olefiant gas, &c. Several metallic alloys might be mixtures of this nature.—Such a combination among the earthy minerals would most readily be effected by the agency of heat, and in such a case it should seem, that the greater the degree of fluidity, the purer (and perhaps less complicated) would be the simple minerals formed upon the cooling of the mass. In all mixtures or solutions of this kind, we might have the ingredients in a variety of proportions, up to the points of saturation for each, and still the form impressed upon the mass as it became solid, would arise from the crystallization of one only.

Without considering the causes which may have brought together the ingredients composing the Bournonite, let us suppose them homogeneously mixed, the question is, whether any of the four Sulphurets enter into a new combination, and thereby form a new integrant molecule, or whether the character is impressed by the crystallization of one in particular.

As it is evident that the number of atoms of the Sulphurets must exactly correspond to those of the Metals, by inspecting the third column of the Table, (page viii. of the Preface,) it is plain that the only combinations which would come within the principles of the Atomic Theory, would be either,

2 Atoms Sulp. Ant. + 2 Ats. Sulp. Lead +
   1 At. Sulp. Cop.
Or, 1 At. Sulp. Ant. + 1 At. Sulp. Lead.

In the first instance we must allow the excess of the Sulphurets of Lead and Antimony, together with all the Sulphuret of Iron, to be mechanically mixed with the essential ingredients. In the second, the whole of the Sulphurets of Copper and Iron, together with the excess of the Sulphuret of Antimony, would form the accidental portion.

A single analysis, however, appears insufficient to decide this question, chemically considered. If the primitive form of the Bournonite should be proved to be distinct from that of the Sulphuret

of Antimony, then there can remain no doubt of
its being entitled to form a distinct species. I
have had no opportunity of investigating this
point, but from the accounts given of these
Minerals by Haüy, Bournon, and Phillips, it does
not appear that their Crystals have been compared,
and as the measurements given of some of their
corresponding angles are nearly identical, it does
not seem improbable that the form of each may
have been impressed by similar integrant molecules.

According to this view of the subject, the
essential composition of an impure Crystal could
never be ascertained from the inspection of a
single analysis, however accurately performed.
Several analyses, undertaken upon crystallized
specimens, selected from different matrices, would
be requisite to point out those ingredients which
were adventitious to the essential character.—
Among the numerous analyses of Augite (from
which have been selected those given in page 88,)
I can find only one in which Magnesia does not
enter as an ingredient, and as there is sometimes
considerable difficulty in separating this Earth
from Lime, it may possibly here have escaped

detection. Omitting, therefore, this analysis as doubtful, I would offer the following conjectures respecting the composition of the Integrant Molecule of this Mineral.—As the primitive form of Diopside is identical with that of Augite, I consider it (from its transparency) as the purest state in which this Mineral is known to occur. From the only analysis hitherto given of it, (see page 87,) the composition of its integrant molecule approaches nearest to a compound of the Bisilicates of Lime and Magnesia, atom to atom. The two analyses of Sahlite (see page 88), the variety of Augite next in purity to the Diopside, would lead to the same result, and even with a much greater degree of probability. Now, with the exception above-mentioned, the three ingredients, Silica, Lime, and Magnesia, are found in all the other analyses of Augite : in that given by Simon, the proportion between these ingredients appears least likely to afford a satisfactory result; but upon examining the proportion between the atoms of each, it does not seem unlikely that there are a sufficient number of integrant molecules of the nature above-mentioned, to effect the crystallization of the whole.

Analysis.	Rep. Nos.	Atoms.
Silica............52.	2.	260
Lime............25.5	3.625	70
Magnesia .... 7.	2.5	28
Ox. Iron 10.5	5.5	19
Alumina 3.5	3.375	10
Ox. Mang.... 2.25	4.5	5
Water 5	1.125	4

It is evident that the number of integrant molecules employed must depend upon the number of atoms of Magnesia, which is the least of the three essential ingredients. Hence, as the atoms of Silica are twice those of the Lime and Magnesia, we shall have 28 Integ. Mols. = 112 atoms Sil. + 28 Lime + 28 Mag.; and therefore there will remain in excess, 148 atoms Sil. + 42 Lime + sum of the atoms of the other substances. All the atoms of Silica in excess might be united with the other ingredients, forming different silicates, but if we suppose them united only to the excess of Lime and Alumina, we have,

42 ats. of Bisil. Lime = 42 Lime + 84 Sil.
10 ats. of Bisil. Alum. = 10 Alum. + 20 Sil.

and therefore the atoms of Silica uncombined, are reduced to 44. Hence, the proportion between the whole of the atoms of the various component molecules would stand thus:

Atoms.

Integrant Molécules,   ⎫
   i. e. Bisil. Lime + Bisil. Mag. ⎬ ... 28

Bisil. Lime 42

Bisil. Alumina 10

Silica 44

Ox. Iron .... ·..............19

Ox. Manganese................ 5

Water 4

Consequently, atoms essential : atoms accidental : :
28 : 124, a greater ratio than 1 : 5, from which it
appears, that the essential atoms bear a greater
proportion to the accidental than in the case of
the Fontainbleau Spar. But in the present instance
we must also recollect, that the accidental ingredients are not all of the same nature, and therefore are not so likely to oppose the crystallization
of those which are essential.—In fact, the atoms
of the essential ingredient are to the atoms of that
accidental ingredient which is in the greatest
excess (viz. Silica) as 28 : 44, which is a greater
ratio than 1 : 2.

In our present state of knowledge, it is impossible to draw any satisfactory arguments from
the comparative magnitude and crystallizing
energies of the integrant molecules of different
minerals ; it may, however, be suggested, that the
dimensions of the more complicated integrant molecules are more likely to be larger than the others,

and would consequently require a smaller number
to produce a crystal of given dimensions, whilst
the latter might easily be included in the cavities
between them.   It is precisely in minerals of this
kind, where the integrant molecule is supposed
to consist of 4 or 5 elementary bodies, that the
greatest confusion prevails—Ex. gr. Garnet, Hornblende, Tourmaline, &c.

It should be remarked, that the excess in the
accidental atoms, in the foregoing example of
the Augite, would be considerably diminished by
every slight addition made to the weight of the
Magnesia (in the original analysis) at the expense
of the Lime; and that if about six parts were
taken from the Lime and added to the Magnesia,
there would no longer be any excess of the atoms
of the former, and the excess of those of the
Silica would be diminished, whilst the number
of the essential atoms would be increased in a
greater ratio than $52 : 28$, that is to say, would
be about doubled.

In all attempts to reduce the analyses of siliceous minerals by the principles of the atomic
theory, the greatest difficulty will be found to
occur among those which contain an alkali, appa-
rently as an essential ingredient.   This substance

almost universally exists (according to the analyses hitherto made) in far too small a quantity to admit of the common adaptation of the theory. Is it likely that some portion may still be retained in combination with the silica? Several analyses might, upon such a supposition, be easily brought within the theory, whilst at present the silica is considerably in excess. But the examples already given have extended this Preface beyond the limits which it was intended to occupy. The chief object in view was to shew, " That when we wish to ascertain the essential ingredients of any crystal, one part of the investigation may be more easily and more correctly deduced, from inspecting the proportion between the absolute *number of atoms* employed in its composition, than between the weights of each ingredient."

In the following Table are given the representative numbers of the Component Molecules of Minerals, which have been adopted in the calculations throughout the Syllabus. They have been selected from various sources, according as each appeared most worthy of insertion.—Some few, which differ from those hitherto published, were calculated from a comparison of different analyses. They are all multiples of .125, the number assumed for hydrogen, not from the idea that this construction (now usually adopted) is strictly true, since it does not appear that the basis upon which such a theory is founded, can be made sufficiently accurate to determine the question ; for, whatever may be the representative number of any substance, as deduced from experiment, if it be not an exact multiple, it must always lie between two multiples of hydrogen, and consequently the difference between this number and one of these multiples cannot exceed half the number assumed for hydrogen; and it will be found, by examining different analyses, that even this difference is generally far too small to create any sensible error in the result. On the present scale, the greatest difference between the number deduced from experiment, and the multiple of hydrogen nearest to it, could not exceed .0625.—As a matter of convenience therefore, and not as a point determined in chemistry, it may be advisable to select the multiples of hydrogen nearest to the numbers given by experiment, and on this account alone has the present Table been so constructed.

ALPHABETICAL LIST

OF THE

COMPONENT MOLECULES OF SIMPLE MINERALS,

With the relative Weights of their Atoms.

Alumina 3.375	Mercury ....25.
Aluminum .... ?	Peroxide of ....27.
Ammonia 2.125	Molybdenum........ 6.
Antimony 5.625	Molybdic Acid ...... 8.
Oxide of....... 6.625	Muriatic Acid 4.625
Arsenic 5.25	
Arsenic Acid....... 7.25	Nickel 3.75
	Oxide of........ 4.75
Barytes 9.75	Nitric Acid 6.75
Bismuth 8.875	
Oxide of....... 9.875	Oxygen .... 1.
Boracic Acid....... 2.75	
Boron............. .75	Palladium ....14. ?
	Phosphoric Acid 3.5
Carbon .... .75	Platinum ....12.5 ?
Carbonic Acid 2.75	Potash .... 6.
Cerium, Oxide of ..12.5 ?	Hydrate of 7.125
Chromic Acid 6.5	
Cobalt........... 5.375	Selenium .... 5.125?
Peroxide of 7.375	Silica .... 2.
Columbium ....,... ?	Silicium............ ?
Oxide of ...... ?	Silver 13.75
Copper .... 8.	Deutoxide of ...15.75
Peroxide of ....10.	Soda .... 4.
	Hydrate of...... 5.125
Fluoric Acid 2.125	Sulphur .... 2.
	Sulphuric Acid ...... 5.
Glucina 3.25	Strontian 6.5
Gold ....16.5 ?	
	Tellurium 4.75
Hydrogen .... .125	Tin 7.375
	Tungstic Acid ......15.
Iridium .... 6. ?	Titanium............ ?
Iron................ 3.5	
Peroxide of 5.5	Uranium...........15.5
Protoxide of 4.5	Oxide of........17.5
Lead ....13.	Water............. 1.125
Peroxide of ....15.	
Lime 3.625	Yttria .... 5.
Magnesia 2.5	Zinc............... 4.375
Manganese 3.5	Oxide of........ 5.375
Protoxide of 4.5	Zirconia 5.625
Mellitic Acid........ ?	

Some of the following Books are among the most useful to those commencing Mineralogy:—

	£.	s.	d.
Allan's Mineralogical Nomenclature, 1 vol. 8vo. ..	0	10	6
Berzelius, de l'Emploi du Chalumeau, 1 vol. 8vo...	0	10	0
Cleveland's Mineralogy, 1 vol. 8vo................	1	4	0
Conversations on Chemistry, 2 vol. 12mo..........	0	14	0
Haüy, Traité de Mineralogie, 2d edit. 4 vol. 8vo. and a 4to. Atlas	3	12	0
Haüy, Traité de Cristallographie, 2 vol. 8vo. and a 4to. Atlas.......................................	2	2	0
Jameson's System of Mineralogy, 2d edit. 3 vol. 8vo.	2	12	6
Phillip's Introduction to Mineralogy, 1 vol. 8vo......	0	12	0

Small Collections of Minerals are fitted up by Mr. MAWE, 149, Strand; and by Mr. G. B. SOWERBY, King Street, Covent Garden.

Wooden Models of Crystals are cut by Mr. N. J. LARKIN, Gee Street, Somers Town, and may be also procured at Mr. MAWE'S.

JOHNSON TUFTS, Downing Street, Cambridge, sells Specimens of various Minerals, Blowpipes, and other mineralogical apparatus.

ERRATA.

—◆—

Page 8, line 4.	For	Light	*read*	No light.
—— 10, —— 6.	——	mamellated	——	mamillated.
——, ——10.	——	85° 45′	——	94° 15′.
——23, ——10.	——	symetrical	——	symmetrical.
————, ——11.	——	10	——	200.
——36, ——17.	——	symetrical	——	symmetrical.
——47, ——12.	——	:: 1000 : 1	——	:: 1 : 1000.
——54, ——13. ⎫				
————, ——19. ⎬	——	symetrical	——	symmetrical.
——55, ——11. ⎭				
——79, ——12.	——	10.6	——	10.5

ERRATA.

A Syllabus

LECTURES ON MINERALOGY.

———◆◆———

Division of Natural History into three branches—
subdivision into two—distinction between or-
ganized and unorganized bodies—Minerals, simple
and compound.

Elementary bodies — compound substances.
Definition of a simple Mineral—integrant molecule
—component (or elementary) molecules.

Ex. gr. Sulphuret of Iron—carbonate of Lime.
Definition of a compound Mineral— *ex. gr.*
Granite.

Distinction between Mineralogy and Geology—
mode of pursuing the study of Mineralogy—want
of a precise arrangement—number of species
inconsiderable.

Method of identifying species—characters of
Minerals—terms used in Mineralogy—subdivision
of Physical and Chemical characters.

B

I. ESSENTIAL CHARACTER.

Composition of the integrant molecule—analysis of Minerals—estimation of the essential and accidental ingredients.

II GEOMETRIC CHARACTERS.

Crystallization—definition of a crystal—to every Mineral a certain series of crystals—*ex. gr.* Quartz— some forms common to different species.

Secondary crystals artificially constructed by the addition of integrant molecules to a primitive nucleus.

Laws of decrement:

    1st. On the Edges—*ex. gr.* Rhomboidal and pentagonal dodecahedrons derived from the Cube.

    2d. On the Angles—*ex. gr.* Regular octohedron derived from the Cube.

    3d. Mixed decrements.

    4th. Intermediate decrements.

Subtractive molecules.

Compound secondary crystals—*ex. gr.* Icosahedron.

Regular pentagonal dodecahedron and regular icosahedron incompatible with the laws of crystallization.

Number of crystalline forms—laws of decrement limited in nature.

Natural mode in which crystals are formed—from solution—from fusion.

Progress of crystallography—opinions of Linnæus—Tournefort—Romé de l'Isle—Bergman—Haüy.

All crystals of the same species reducible to one form—cleavage—*ex. gr.* Rhomboid extracted from all crystals of carbonate of lime—some primitive forms common to different species—Structure—Fracture.

Five kinds of primitive forms:

   1. Regular tetrahedron.
   2. Parallelopipeds.

(a) Cube.

(b) Rhomboid.

(c) Quadrangular prisms.

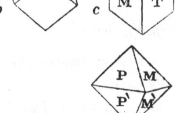

   3. Octohedrons.

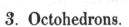

   4. Regular hexahedral prism.
   5. Rhomboidal dodecahedron.

Forms of some integrant molecules differ from

those of the primitive crystals—three kinds of integrant molecules:

1. Tetrahedrons.
2. Triangular prisms.
3. Parallelopipeds.

III. EXTERNAL CHARACTERS.

Regular crystals—their form supposed to be independent of the relative magnitude of their faces—method of determining their form by the common goniometer—reflecting goniometer.

Hemitrope crystals.

Indeterminate crystallization—*ex. gr.* lenticular—acicular—striated—fibrous.

Concretions—stalactites and stalagmites—incrustations—petrefactions—pseudomorphous crystals—amorphous masses.

IV. PHYSICAL CHARACTERS.

Specific gravity $\left\{ = \frac{\text{weight}}{\text{wt. lost in water}} \right\}$ estimated by the common scales—Hydrostatic balance—Nicholson's balance.

Cohesion of the Particles—Hardness, estimated by the file—point of the knife—mutual action of substances—percussion with steel wheel—Frangibility—difference between fragile and tender bodies.

Flexibility—Elasticity—Malleability—Ductility.

Action of light—Transparency—translucency
—opacity.

Colour—from reflection and refraction—essential and accidental—chatoyant—irised—scratch—streak—powder.

Double refraction—Polarization of light—Phosphorescence—produced by heat—friction—after exposure to light.

Touch—adhesion to the tongue.

Odour—produced by breathing, friction, and heat.

Taste.

Magnetism—with and without polarity—double magnetism.

Electricity—positive and negative; or vitreous and resinous—excited by friction—heat—pressure.

V. Chemical Characters.

Action of Heat—excited by the common blowpipe—hydraulic blowpipe—gas blowpipes.

" Apparatus for the common blowpipe :

" *Supports*—Charcoal from pine-wood or willow
" —Platina-foil, wire, and forceps—glass-tubes—
" bone-ash and tobacco-pipes.

" *Re-agents*—Carbonate of soda—borax—glass
" of borax—phosphate of soda and ammonia—
" nitre."

Action of water and the atmosphere—
*Action of acids—*Nitric—muriatic—sulphuric—
Action of alkalis—
Alloys.

VI. EXTRANEOUS CHARACTERS.

VII. DISTINCTIVE CHARACTERS.

VIII. GEOLOGICAL AND GEOGRAPHICAL POSITIONS.

IX. USES AND OBSERVATIONS.

HYDROGEN—OXYGEN—WATER—
OXIDES — ACIDS — ALKALIS — EARTHS.

GENUS I. Sulphur.

S. G. = 1.9907—Purification—Artificially crystallized from fusion or solution—casts.

Species 1. Native Sulphur.

II. *Prim.* rhomboidal octohedron $P.P' = 143^\circ 7'$.

III. Crystals 6. Amorphous—disseminated—pulverulent.

IV. S. G. = 2.0332.

VIII. Volcanic and other native sulphur compared.

Species 2. Sulphuric Acid.

Its extensive uses in chemical operations and the arts.

GENUS II. Carbon.

Charcoal—uses in Chemistry—metallurgy—steel—printing ink—fuel.

Species 1.   Diamond.

II. *Prim.* regular octohedron.

III. Crystals.

IV. S. G. = 3.5 to 3.6—Light refracted if the angle of incidence is greater than $24\frac{10}{2}°$.

IX. Art of polishing Diamonds—Diamond powder —the cutting Diamond.

APPENDIX I.

Mineral Carbon.

Plumbago— *Graphite—Black-lead.*

I. Carbon 90.9 + Iron 9.1.   Berthollet.

III. Amorphous—disseminated.

IV. S. G. = 2.08 to 2.32.

IX. Black-lead pencils—recent and artificial formations of Plumbago.

Anthracite— *Blind Coal—Stone Coal.*

I. Carbon....72.05 ⎫
   Silica ....13.19 ⎪ Phillips.
   Alumina .. 3.29 ⎬
   Ox. Iron .. 3.47 ⎭

III. Amorphous—slaty—columnar.

IV. S. G. = 1.4 to 1.8.

APPENDIX II.

Carburetted Hydrogen.

Mineral Oils.

    (a) Naptha.

	Sauss.	Theory.		
I. Carbon....87.21	10.47		
————....——	76.74	1	
Hydrogen..12.79	12.79	1	

(*b*) Petroleum.

Mineral Bitumens.

(*a*) Earthy — *Maltha.*

(*b*) Elastic — *Mineral Caoutchouc.*

(*c*) Compact — *Asphalt.*

S. G. = 1. to 1. 6.

Mineral Resins.

(*a*) Amber — *Electrum.*

III. Amorphous — Stalactitic.

IV. S. G. = 1.078 to 1.085.

(*b*) Fossil Copal — *Highgate Resin.*

III. Amorphous.

IV. S. G. = 1.046.

(*c*) Retinasphalt.

Coals.

(*a*) Jet.

(*b*) Cannel Coal — 39 to 75 per cent. of Carbon, with Bituminous oil and earth.

IV. S.G. = 1.2.

(*c*) Common Coal — *Black Coal.*

I. Carbon 56 . 8 + Asphalt and Maltha 43.

IV. S. G. = 1 . 2 to 1 . 3.

(*d*) Lignite — *Brown Coal.*

Appendix III.

Carbonic Acid.

GENUS III. Boron.

Species 1. Boracic Acid.

	Gmelin.	Berz.	Theory.	
Boron....	74.4	74.17	72.5+	1
Oxygen ..	25.6	25.83	27.4+	2

III. Mamellated incrustations.

IV. S. G. $= 1.48$.

GENUS IV. Silicium.

Species 1. Silica—*Quartz.*

II. *Prim.* Obtuse rhomboid P. P$'$ $= 85^\circ 45'$.

III. Crystals 13. Acicular—fibrous—granular—arenaceous—concretionary — stalactitic — pseudomorphous—amorphous.

Rock Crystal—Amethyst, 97.5 per cent. of Silica —*Milk* — *Rose* — *Smoky*—*Fat*—*Avanturine*— *irisated—ferruginous—Elastic.*

Hyalite.. $\left\{ \begin{array}{l} \text{Silica ..92} \\ \text{Water.. 6.33} \\ \text{Loss .. 1.66} \end{array} \right\}$ Bucholtz.

IV. S. G. $= 2.0499$ to 2.816.

VI. Cat's-eye.

Chalcedony.

Silica 83	84		
Alumina...... 2	16	}	Berzelius.
Lime 11	..		

Onyx—Sard — Sardonyx —Cachalong—Carnelian —Plasma—Prase—Crysoprase — Heliotrope or Bloodstone.

Agate—Ribbon—Brecciated—Fortification—Moss —Mocha-stone.

Flint— Chert — Hornstone — Swimming Flint — Siliceous Sinter.

Opal —- precious — Hydrophane — Semi-Opal — Wood Opal—Ferruginous—Menilite.

Jasper—43 to 80 per cent. of Silica—Common— Jasper-Opal—Ribbon—Egyptian Pebble—Por-cellanite.

GENUS V. Arsenic.

Dyes and Paints—Artificial gems.

Species 1. Native Arsenic.

ii. *Prim.* Regular octohedron.

iii. Amorphous—Concretionary.

iv· S.G. = 5.67 to 5.7633.

Species 2. Oxide of Arsenic.

	Proust.	Theory.		
i. Arsenic......75		72.41+	1	
Oxygen......25		27.58+	2	

ii. *Prim.* Regular octohedron.

iii. Crystal 1. Acicular—pulverulent.

iv. S. G. = 3.706 to 5.

Of Arsenic Acid.

Species 3. Sulphuret of Arsenic—*Realgar*.

	Then.	Klap.	Theory.	
i. Arsenic	75	69	72.41+	1
Sulphur	25	31	27.58+	1

ii. *Prim.* Oblique rhomboidal prism.

iii. Crystals 3. Acicular — concretionary — dis-seminated—amorphous.

iv. S. G. = 3.3384 to 3.35.

ix. Its uses among the Chinese—Paint.

Species 4. Bi-sulphuret of Arsenic?—*Orpiment.*

	Klap.	Then.	Theory.	
i. Arsenic	62	57	56.75+	1
Sulphur	38	43	43.24+	2

iii. Laminar—disseminated—amorphous.

iv. S. G. = 3.35 to 3.4522.

ix. Uses among the Turks—Paint—preparation of artificial orpiment.

GENUS VI. MOLYBDENUM.

S. G. = 7. to 8.611.—Molybdic Acid.

Species 1.   Sulphuret of Molybdenum.

	Pell.	Seybert.	Buch.	& Theory.
I. Molybdenum....45	59.42	60	1
Sulphur........55	39.68	40	2
Loss	.9	..	

III. Crystals 2.   Amorphous.
IV. S. G. = 4.5 to 4.74.

TUNGSTEN and Tungstic Acid.
CHROME and Chromic Acid.

GENUS VII.   IRON.

S. G. = 7.788—Its great abundance and numerous uses—steel.

Species 1.   Native Iron.

I. Iron 92.5 to 94 per cent.
III. Amorphous—plated, ramose, cellular.
VIII. Volcanic and Meteoric.
      Meteorites—Alloy of Iron 98.5 + Nickel 1.5.
      S. G. = 6.48 to 7.4.

Species 2.   Peroxide of Iron.   *Specular Iron ore—fer oligiste.*

	Hass.	Theory.	
I. Iron 69	63.63 +	1
Oxygen........31	36.36 +	2

II. *Prim.* Acute rhomboid $P.P' = 86^0 \ 10'$.

III. Crystals 6—Lamelliform.

IV. S. G. = 5.

Species 3.   Oxydulated Iron—*Magnetic Iron Ore.*

	Berz.	Theory.	
I. Protox. Iron....	28.14	29.0+	1
Perox. Iron	71.86	70.9+	2

Buch.

Iron......70 } or { Protox. Iron..49.5
Oxyg......29 }   { Perox. Iron ..49.5

II. *Prim.* Regular octohedron.

III. Crystals 3.   Lamelliform — Compact *(native loadstone)*—Earthy.

IV. S. G. = 4.4.

VI. Titaniferous—S.G. = 4.6 to 4.9.

Oxide Titan. .. 8 to 15.9 } per cent.
Oxide Iron....79 to 98.7 }

Chromiferous — *Chromiron* — Chrome 2 per cent.

Appendix I.   Red Iron Ore.

Var. 1.   Fibrous—*Red Hæmatite.*

I. Ox. Iron 90 per cent.

III. Botryoidal—Stalactitic.

IV. S. G. = 4.7 to 5.

Var. 2.   Compact.

III. Cubic (pseudomorphous?)—slaty—amorphous.

IV. S. G. = 3.5 to 5.

Var. 3.   Scaly.

i. Iron 66.0 + Oxygen 28.5 + Sil. and Al. 5.5
   Henry.

Var. 4. 5.   Red Ochre—Red Chalk, (Reddle.)

APPENDIX II.   Brown Iron Ore.

Var. 1.   Cubic (pseudomorphous?)

Var. 2.   Fibrous—Brown Hæmatite.

	Vauq.	Theory.		
i. Peroxide Iron	80.25	73.3 +	1	
Water	15.	15.	1	
Silex	3.75			

iii. Mamillary—stalactitic.
iv. S.G. = 3.7 to 4.

Var. 3.   Compact.   Peroxide Iron 84 per cent.
iv. S.G. = 3.5 to 3.7.

Var. 4.   Scaly.   Var. 5.   Ochreous.

Var. 6.   Umber—Ox. Iron 48, Ox. Mang. 20,
   per cent.

APPENDIX III.   Black Iron Ore.

Var. 1.   Fibrous—Black Hæmatite.
iv. S.G. = 4.7.

Var. 2. Compact.   Var. 3. Ochreous.

APPENDIX IV.

Jaspery Iron Ore — Clay Iron Stone — Bog Iron
   Ore — Pitchy Iron Ore.

Species 4. Sulphuret of Iron. *Magnetic Iron Pyrites.*

<table>
<tr><td></td><td>Hatchet.</td><td>Theory.</td><td></td></tr>
<tr><td>ɪ. Iron........</td><td>63.5</td><td>63.63</td><td>1</td></tr>
<tr><td>Sulphur ....</td><td>36.5</td><td>36.36</td><td>1</td></tr>
</table>

ɪɪ. *Prim.* Right rhomboidal prism.

ɪɪɪ. Crystals—amorphous.

ɪv. S. G. = 4.51.

Species 5. Bi-sulphuret of Iron. *Iron pyrites.*

<table>
<tr><td></td><td colspan="4">Hatchet.</td><td>Theory.</td><td></td></tr>
<tr><td>ɪ. Iron ....</td><td>47.85</td><td>47.5</td><td>47.3</td><td>46.4</td><td>46.66</td><td>1</td></tr>
<tr><td>Sulphur..</td><td>52.15</td><td>52.</td><td>52.7</td><td>53.6</td><td>53.33</td><td>2</td></tr>
</table>

<table>
<tr><td></td><td>Bucholtz</td><td>Theory.</td><td>Proust</td><td>Theory.</td><td></td></tr>
<tr><td>Iron ....</td><td>44.85</td><td>44.75</td><td>47.36</td><td>46.06</td><td>1</td></tr>
<tr><td>Sulphur..</td><td>51.15</td><td>51.15</td><td>52.64</td><td>52.64</td><td>2</td></tr>
</table>

ɪɪ. *Prim.* Cube.

ɪɪɪ. Crystals 40. Amorphous—pseudomorphous.

ɪv. S. G. = 4.6 to 4.8.

vɪ. White Iron pyrites.

<table>
<tr><td></td><td>Hatchet</td><td>Theory.</td></tr>
<tr><td>Iron........</td><td>45.66</td><td>45.66</td></tr>
<tr><td>Sulphur ....</td><td>54.34</td><td>52.18+</td></tr>
</table>

Arsenical Iron pyrites—Auriferous.

Species 6.   Sulphate of Iron—*Green vitriol.*

	Berg.	Theory.	
I. Iron........	23	26.9+	1
Sul. Acid	39	38.4+	1
Water	38	34.6+	4

II. *Prim.* Acute rhomboid P. P. = 81° 23′.

III. Crystals 7—fibrous—pulverulent—amorphous —stalactitic.

Species 7.   Carbonate of Iron—*Spathose Iron —Brown Spar.*

	Bert.	Theory.	Bayen	Theory.	
I. Protox. Iron ..	57.	57.	66	55.6	1
Carb. Acid	35.	34.8+	34	34.	1
Manganese....	1.5	&c.			
Magnesia	5.5				
Loss	1.				

II. *Prim.* Obtuse rhomboid P. P′. = 107°.

III. Crystals—fibrous—amorphous.

IV. S. G. = 3.6 to 3.8.

Species 8.   Phosphate of Iron.

	Laugier	Theory.		Cadet	Theory.	
I. Iron	41.25	41.25	1	41.1	41.1	1
Phos.Acid	19.25	32.08⎞	1	36.9	31.9⎞	1
Water ..	31.25	10.31 ⎬	1	13.1	10.2⎬	1
....		8.11⎠		....	7.8⎠	
Silica	1.25	&c.		3.	&c.	
Alumina..	5...			5.8		
Lime	...			9.1		

	Bert.	Theory.	Vauq.	Theory.	Klap	Theory.
Iron ......43.	43		31	31	47.2	47.2
Phos. Acid 23.1	33.4⎞		27	24.1	32.	36.7⎞
Water 34.4	10.7⎬		..	7.7	20.	11.8⎬
....	13.3⎠		..	...	..	4.5⎠
Mang....... ..3	...		42	...		

II. *Prim.* Oblique rectangular prism.

III. Crystals 12. Earthy.

IV. S. G. = 2.697.

Species 9. Silicate of Iron.—*Yenite.*

	Hising.	Theory.	Vauq.	Theory.	
I. Iron ......55.	55.		57.5	57.5	1
Silica ....28.	23.3+		30.	25.5+	1
....	4.6+		..	4.4+	
Lime ....12.	&c.		12.5	&c.	
Alum. 6			..		
Mang..... 3.			3.		
Loss...... 1.4			..		

II. *Prim.* Rectangular octohedron P.P'. = 113° 2'.

III. Crystals 6.—Acicular—amorphous.

IV. S. G. = 3.8.

Species 10. Arsenical Iron.—*Mispickel.*

	Phill.	Theory?		Vauq.	Strom.	Thom.
I. Arsenic .42.1	42.85+		1	53.	42.88	48.1
Iron ...58 9	57.14+		2	19.7	36.04	36.5
Sulphur	....			15.3	21.08	15.4

II. *Prim.* Right rhomboidal prism. M. T. =
111° 18′.

III. Crystals 40.—Amorphous.

IV. S. G. = 5.6 to 6.5223.

Species 11. Arseniate of Iron.

	Chenevix	Vauq.
I. Ox. Iron ..45.5	48	
Ars. Acid ..31.	18	
Water 10.5	32	
Silica .... 4.	..	
Lime....... ..	2	
Copper.... 9.	..	

II. *Prim.* Cube.

III. Crystals 4.—Concretionary—amorphous.

IV. S. G. = 3.

Species 12. Chromate of Iron.

	Tassaert	Vauq.	Laug.	Klap.	
I. Ox. Iron ....36.		34.7	34.	33.6	
Chromic Acid 63.6		43.	53.	55.5	
Silica		2.	1.	2.	
Alumina		20.3	11.	6.	
Loss 4		..	1.	3.5	
Ox. Iron ....40.9 +		29.7	34.	33.6	1 ⎰ Theory
Chrom. Acid 59.0 +		43.	49.1	36.3	1 ⎱

II. *Prim.* Regular octohedron.

III. Octohedron?—granular—amorphous.

IV. S. G. = 4.03.

Species 13.   Tungstate of Iron.—*Wolfram.*

	Berzel.	Elhuy.	Vauq.	Klap.
I. Ox. Iron	18.32	13.5	18.	31.2
Tung. Acid....	78.77	65.	67.	46.9
Mang	6.22	22.	6.25	..
Silica.........	1.25	2.	1.5	..
Loss	..	..	7.25	21.9
Ox Iron	18.32	13.5	18	1 } Theory.
Tung. Acid....	61.06	45.	60	1 }

II. *Prim.* Right rectangular prism.

III. Crystals 3.—Amorphous.

IV. S. G. = 7.1 to 7.3.

GENUS VIII.   NICKEL.

S. G. = 8.5 to 8.93.   Plates $\frac{1}{100}$ inch thick—Magnetic power of Nickel : Iron :: 1 : 4 nearly—or as 35 : 55 (Lampadius).

Species 1.   Native Nickel.

III. Capillary.

Species 2.   Arsenical Nickel—*Copper Nickel.*

	Strom.	Theory.	
I. Nickel 44	41.25	1	
Arsenic 55	57.75	1	
Iron			
Lead ....} 1			
Sulphur ..)			

III. Six-sided prism? — reticulated — dendritic — concretionary—amorphous.

IV. S. G. = 6.6086 to 7.56.

VI. Argentiferous.

Species 3. Arseniate of Nickel?—*Nickel Ochre.*

	Phill.	Klap.
I. Ox. Nickel....37	15.62
Arsenious Acid .37	....
Water........24	....
Ox. Iron 1	4.58
Sulp^c. Acid 1	....
Silica	35.
Alumina	5.
Magnesia	1.25
Lime	.4
Loss .... ⎱	37.
Arsenic Acid ?.. ⎰	....

III. Crystals—pulverulent—amorphous.

Species 4? Black Ore of Nickel.

GENUS IX. MANGANESE.

Species 1. Oxide of Manganese.

II. *Prim.* Right rhomboidal prism M. T. = 100°.

III. Crystals 6.—Acicular—earthy *(Wad.)*—amorphous.

IV. S. G. = 3.7076 to 4.756.

vi. Siliciferous—white, rose, and violet.

ix. For clarifying and tinging glass—used for obtaining oxygen.

Species 2. Sulphuret of Manganese.

	Vauq.	Klap.
i. Mang........	85	82
Sulph........	15	11

iii. Amorphous.

iv. S. G. = 3.49 to 3.95.

Species 3. Carbonate of Manganese.

	Descot.	Lamp.
i. Ox. Mang.....	53.	48.
Carb. Acid	36 6	49.2
Ox. Iron	8.	2.1
Silex	..	9.

iii. Amorphous.

iv. S. G. = 2.8.

Species 4   Phosphate of Manganese.

	Vauq.
i. Ox. Mang.....	42
Phos Acid	27
Ox. Iron	31

iii. Amorphous.

GENUS X.   TIN.

S. $\overset{\prime}{G}$ = 7.291 to 7.299—Wire $\frac{1}{10}$ inch supports 49.5 lbs.—tin foil—coating for iron plates.

Species 1.   Per-oxide of Tin.

	Klap.	Theory.	Klap.	Theory.	
I. Tin 77.5	77.5	75.	75.	1	
Oxygen.21.5	21.0	23.75	20.3	2	
Iron .. .25	&c.	.5	&c.		
Silex .. .75		.75			

II. *Prim.* Symetrical octohedron.
III. Crystals 10—fibrous *(wood tin)*—concretionary
     —granular *(stream tin)*—amorphous.
IV. S.G. = 6.76 to 6.95.

Species 2.   Sulphuret of Tin?

	Klapr.	
I. Tin........26.5	34.	
Sulph.......30.5	25.	
Iron ....12.	2.	
Copper .....30.	36.	

III. Disseminated—amorphous.
IV. S. G. = 4.35 to 4.785.

GENUS XI. Copper.

S. G. = 7.78 to 7.87 and 9. Wire $\frac{1}{10}$ inch supports 299 lbs.— Bronze — Bell-metal — Sheet-copper — 10844 Tons raised in Great Britain and Ireland in one year ending 30 June 1822.

Species 1. Native Copper.

II. *Prim.* Cube or regular octohedron.

III. Crystals 100.—Capillary — dendritic — amorphous.

IV. S. G. = 7.6 to 8.5844.

Species 2. Protoxide of Copper.—*Red oxide of Copper—Ruby Copper.*

	Klap.	Chen.	Theory.	
I. Copper......91.	88.5	88.8+	1	
Oxygen .... 9.	11.5	11.1+	1	

II. *Prim.* Regular octohedron.

III. Crystals 100—capillary—amorphous.

IV, S. G. = 3.95 to 5.691.

VI. Ferruginous—*Tile ore.*
    S. G. = 3.572.—Copper = 10 to 50 per cent.

Species 3. Per-oxide of Copper?—*Black Copper.*

III. Disseminated—reniform and fibrous.

Species 4. Sulphuret of Copper.—*Vitreous copper—copper glance.*

	Thom.	Klap.	Klap.	Theory.	
I. Copper ......73.	76.5	78.5	77.6	1
Sulph. 24.5	22.	18.5	19.4	1
Iron .... 1.	.5	2.25	&c.	
Silica 1.8	..	.75		
Loss 2	1.	..		

Or,

Sulph. Cop....60.	80.
Bi-sul. C......37.5	18.

II. *Prim.* Regular hexahedral prism.

III. Crystals 120.—Amorphous.

IV. S. G. = 4.81 to 5.452.

VI. Variegated vitreous copper.

Purple copper.—*Buntkupfererz.*

	Klap.	Theory.		Klap.	Gueniv.	
Copper....69.5	38.	1	58	74.5	47.
Sulph. ....19.	19.	1	19	20.5	13.
Iron 7.5	&c.		18	1.5	9.3
Oxygen .. 4.			5	..	25. Silex
Loss....... ..			..	3.5	7. lime

Species 5. Bisulphuret of Copper.—*Copper Pyrites.*

	Chen.	Lamp.	Gueniv.
I. Copper ......30.	41.	30.
Sulphur ......12.	45.1	36.5
Iron	17.1	31.
Zinc	..	1.
Ox. Iron ......53.	..	..
Silica .... 5.	..	1.
Loss	..	...5
	100.	103.1	100.

Or,

	Chen.	Lamp.	Gueniv.
1. Copper ......28.⎱	41. ⎱	30. ⎱
2. Sulp. ....14.⎰	20.5⎰	15. ⎰
1. Iron	17.1⎱	5.7+⎱
2. Sulp.	19.5⎰	6.5+⎰
1. Iron	..	25.2+⎱
1. Sulp.	..	14.4+⎰
1. Zinc	..	1. ⎱
1. Sulp.	..	.4+⎰
&c. ....58.	..	14.5
	100.	98.1	99.7

II. *Prim.* Regular tetrahedron.

III. Crystals—concretionary—amorphous.

IV. S. G. = 4.3154.

IX. Mode of extracting the copper.—In Cornwall, (1817), 73727 tons of ore yielded 6425 tons of copper.

APPENDIX.

(a) Grey Copper Ore—*Fahlerz.*

	Klaproth			Thoms.
Copper ......41.	48.	42 5	19.2
Iron 22.5	25.5	27.5	51.
Arsenic 24.1	14.	15.6	15.7
Antimony	..	1.5	..
Silver 4	.5	.9	..
Sulphur ......10.	10.	10.	14.1
Loss .... 2.	2.	2.	..

(b) Black Copper Ore—*Graugiltigerz.*

	Klaproth			
Copper ...31.36	37.75	39.	40.25	26.
Iron 3.3	3.25	7.5	13.5	7.
Arsenic	..	..	.75	..
Antimony ..34.09	22.	19.5	23.	27.
Silver 14.77	.25	..	.3	13.25
Sulphur ..11.5	28.	26.	18.5	25.5
Zinc	5.	..	..	..
Mercury	..	6.25	..	..
Alumina .. .3	..	..	..	..
Loss 4.68	3.75	1.75	3.7	1.25

(c) Tennantite.

	R. Phill.
Copper 45.32
Arsenic ...,11.84
Iron 9.26
Sulphur ,....28.74
Silica ...... 5.

Species 6.   Sulphate of Copper.—*Blue Vitriol.*

	Berzelius	Theory.	
I. Per-ox. Copper ..	32.13	32.14	1
Sulph. Acid	31.57	32.14	2
Water	36.30	36.16	10
	100.	100.44	

II. *Prim.* Irregular oblique four-sided prism. P. M. = 109° 32′, M. T. = 124° 2′

III. Crystals 11.—Concretionary—stalactitic—pulverulent—amorphous.

Species 7.   Anhydrous Carbonate of Copper.

	Thoms.	Theory.	
I. Per-ox. Copper ..	60.75	60.74	1
Carbonic Acid....	16.7	16.71	1
Per. ox. Iron	19.5	&c.	
Silica	2.1		
Loss	.95		

IV. S. G. = 2.62.

Species 8.   Green Carbonate of Copper—*Malachite.*

	Fontana	Phill.	Theory.	
I. Per-ox. Copper ..	75.	72.2	72.07	1
Water	5.6	9.3	8.10	1
Carb. Acid	19.4	18.5	19.81	1

	Klap.	Proust.	Vauq.	Theory.	
Copper	58.	56.8	56.1	57.65	1
Oxygen ..	12.5	14.2	14.	14.41	2
Water	11.5	..	8.65	8.10	1
Carb. Acid	18.	27.	21.25	19.81	1
Silica	1.	..	..		
Lime	1.	..	..		

III. Capillary — fibrous — stalactitic — pseudomorphous—amorphous.

IV. S. G. = 3.5.

Species 9. Blue Carbonate of Copper.

	Phillips	Theory.	
I. Per-ox. Copper	69.08	69.36	3
Carb. Acid	25.46	25.43	4
Water	5.46	5.20	2

Or ?

Green Carb. Copp.	64.14	2
Bi. Carb. Copp.	35.83	1

	Vauq.	Klap.	Theory.		Pell.
Copper	56.	56.	55.26	3	68
Oxygen	12.5	14.	13.81	6	9
Carb. Acid	25.	24.	25.43	4	19
Water	6.5	6.	5.20	2	2

II. *Prim.* Irregular octohedron.

III. Crystals—concretionary—earthy.

IV. S. G. = 3.2 to 3.6082.

IX. Turquoise—Armenian stone.

Species 10. Muriate of Copper.

	Berth.	Proust.	
I. Copper	52.	46.8	57.4
Oxygen	11.	11.5	14.6
Mur. Acid	10.	9.5	10.
Water	12.	15.	12.
Silica	11.	17.	..
Lime	..	..	4.
Iron	1.	..	2.

III. Pulverulent.

Species 11.   Phosphuret of Copper?

Species 12.   Phosphate of Copper.

	Klap.	Theory.	
I. Per-oxide Copper....	68.13	68.37	1
Phosphoric Acid	30.95	23.93	1
Loss (Water?)........	.92	7.69	1

	Lunn	Theory.		Quere?	
Per-ox. Cop	18.1	62.847+	1	18.1	68.31+
Phos. Acid	6.246	21.687+	1	6.246	23.56+
Water (&	2.15	15.454+ }	2	2.15	8.11+
Loss).... }	2.304	.... }		....	....
	28.8	99.988+		26.496	99.98+

II. *Prim.* Rectangular octohedron.

III. Crystals 3—fibrous—amorphous.

IV. S. G. = 3.5 to 4.07.

Species 13.   Arseniate of Copper.

I.		Chenevix.					Vauq.		Klap.
Per-ox.C.	49	60.	54	50	51	58	59	39	50.62
Ars. Acid	14	39.7	30	29	29	21	41	43	45.
Water	35	....	16	21	18	21	....	17	3.8
Loss........	....	....	....	....	2	....	....	1	.88
Var.	1	4	3	5	6	2	4	2	4

III.   *Var.* 1. Obtuse octohedral—S. G. = 2.88.

     Var. 2. Hexahedral—S. G. = 2.5.

     Var. 3. Trihedral—S. G. = 4.2.

     Var. 4. Prismatic—S. G. = 4.2—Capillary.

     Var. 5. Hæmatitic—*Wood copper.*

     Var. 6. Amianthiform.

Species 14.   Martial Arseniate of Copper.

	Chenev.
I. Ox. Copper	....22.5
Arsenic Acid	... 33.5
Ox. Iron	.... 27.5
Water	..........12.
Silica	.... 3.

III. Crystals 2.

Species 15.   Bisilicate of Copper?—*Dioptase—Crysocolla—Emerald copper.*

I.	Klap.	Theory?	Lowitz	Theory?	John.	
Copper	40	40	1	}55	54.7+	1	42.	37.8
Oxygen	10	10	2				7.63	8.
Silica	26	20	2	33	32.8+	3	28.37	29.
Car. Acid	7	&ᶜ		....	....		3.	....
Water	17	....		12	12.3+	2	17.5	21.8
Sulp. Lime	....	....		....	. ..		1.5	3.

II. *Prim.* Obtuse rhomboid P. P′. = 126° 17′.

III. Crystals.

IV. S. G. = 3.3.

Species 16.   Seleniuret of Copper?

GENUS XII. LEAD.

S. G. = 11.352—Minium—Massicot—Litharge—Ceruse—Use in glass.

Species 1. Native Lead.

Species 2. Protoxide of Lead?—*Native massicot.*

	John.	Theory.
I. Lead..............	82.6923	Protox. Lead 30.486
Oxygen	10.5768	Per-ox. Lead 41.803
Carb. Acid ...	3.8462	Carb. Lead ... 24.825
Ox. Iron and Lime }	.4808	&c......... ..... 2.884
Copper...........	trace	
Ferruginous Silica }	2.4039	

III. Amorphous.
IV. S. G. = 8.

Species 3. Per-oxide of Lead—*Native Minium.*

III. Pulverulent—amorphous.

Species 4. Sulphuret of Lead—*Galena.*

I.	Vauq.	Theory.	Westr.	Theory.	Thoms.	Theory.	
Lead. 54	53.7+	83.00	86.18+	85.13	85.08+	1	
Sulph. 8	8.2+	16.41	13.22+	13.02	13.06+	1	
Silic. ⎱38 Lime ⎰	&c.	&c.		&c.			

			Vauquelin.			Meade.
Lead....69.	69.	68.	64.	72.
Sulphur .16.	18.	16.	18.	24.
Lime & ⎱15. Silica ..⎰	13.	16.	18.	4.

							Theory.
Lead....65.	66.5+	64.2+	62.9+	73.4+	1	
Sulphur .20.	20.4+	19.7+	19.0+	22.5+	2	

II. *Prim.* Cube.

III. Crystals—granular—amorphous.
Specular *(Slickensides.)*

IV. S. G. = 7.5873.

VI. Antimoniferous—16 to 28 per cent. of Antimony.
Argentiferous—$11\frac{1}{2}$ oz. to 135. oz. of silver per ton.
Ferriferous.
Supersulphuret.
Blue-Lead; pseudomorphous; S. G. = 5.4.

IX. Usual ore of lead—glaze for pottery.

Species 5.   Sulphate of Lead.

	Klapr.	Theory.	Klapr.	Theory.	
I. Ox. Lead..	71.	71.	70.5	70.5	1
Sulp. Acid .	24.8	23.6+	25.75	23.5	1
Water	2.	&c.	2.25	&c.	
Ox. Iron ..	1.				
Loss......	1.2				

II. *Prim.* Rectangular octohedron.

III. Crystals 8—granular—amorphous.

IV. S. G. = 6.3.

Species 6.   Carbonate of Lead.

	Westrumb.		Theory.	Lampad.	Theory.	
I.						
Ox. Lead..	80.25	81.2	81 2	78.5	78.5	1
Carb. Acid.	16.	16.	15.72	18.	14.3	1
Ox. Iron ..	.18	.3	&c.	..	&c.	
Alumina ..	.75	..		..		
Lime	.5	.9		..		
Water	..	..		2.		
Charcoal	..	..		1.5		
Loss	2.32	1.6		..		

II. *Prim.* Rectangular octohedron P. P′. = 108° 20′.
    M. M′. = 117° 18′.

III. Crystals—acicular—earthy—amorphous.

IV. S. G. = 6.0717 to 6.72.

VI. Cupriferous.

Species 7.   Muriate of Lead—*Murio-carbonate of Lead.*

	Klap.	Theory.		Klap.	Chen.
I. Ox. Lead ..55	61.5+	1		85.5	85.
Mur. Acid..45	38.4+	2		8.5	8.
Carb. Acid ..				6.	6.
Loss........				..	1.

III. Crystals 4.

IV. S. G. = 6.0651.

Species 8.   Phosphate of Lead.

I.	Klap.	Theory.	Four.	Klap.	Vauq.	Laugier.
Ox. Lead ..80.	79.4+	1	79	78.4	49.23	76.8
Phos. Acid..18.	18.5+	1	18	18.37	18.17	9.
Ox. Iron......	&c.		1	1.	..	1.5
Mur. Acid.. 1.6			..	1.7	..	4. ars.
Water........			2	..	..	7.
Silica........				..	32.	..
Loss.........				..	..	1.7

II. *Prim.* Obtuse rhomboid P. P'. $= 110^\circ 55'$.

III. Crystals—mamillary—amorphous.

IV. S. G. = 6.4 to 6.9411.

VI. Arseniated.

Species 9.   Arseniate of Lead.

	Gregor.	Theory.		Rose.
I. Ars. Acid ... 26.4	26.4		1	19.05
Ox. Lead 69.76	54.61+		1	73.13
....	5.11+ }			&c.
Mur. Acid .. 1.58	1.58 }			
Ox. Iron .. }	2.26	&c.		
Silica, &c... }				

iii. Crystals—capillary—mamillary—amorphous.

iv. S. G. = 5.0466 to 6.4.

vi. Reniform—S. G. = 3.9.

Bindheim	Theory.
Ox. Lead....25.	Arseniate of Lead 37.0 +
Ox. Iron ....14.	Arseniate of Iron 22.6 +
Ars. Acid....25.	Perox. Iron............ 4.2 +
Silver 1.15	&c......................
Silica .... 7.	
Alumina .... 2.	
Water........10.	

Species 10.　Molybdate of Lead.

I.	Klap.	Theory.	Macq.	Theory.		Hatch.
Ox. Lead 64.62		64.35	58.75	56.57 +	1	58.4
Molyb. Acid 34.25		34.32	28.	30.17 +	1	38.
Loss 1.33		&c.	&c.			&c.

ii. *Prim.* Symetrical octohedron P. P'. = 76° 40'.

iii. Crystals 10—amorphous.

iv. S. G. = 5.09 to 5.48.

Species 11.　Chromate of Lead.

I.	Vauq.		Then.	Theory.	
Ox. Lead....63.96	65.1	64.	69.7 +	1	
Chr. Acid....36.4	34.9	36.	30 2 +	1	

ii. *Prim.* Oblique rhomboidal prism.

iii. Crystals—lamelliform.

iv. S. G. = 5.75 to 6.

Species 12.　Chromite of Lead?

GENUS XIII.　Zinc.

S. G. = 7.19.—Malleable and ductile, between 210^d and 300^d Fahr.—Flowers of Zinc—Brass—Princes metal—Pinchbeck—Tombac.

Species 1.　Oxide of Zinc—*Red Zinc Ore.*

	Bruce.	Theory.		
i. Zinc 76	74.7 +	1
Oxygen 16	17.2 +	1
Ox. Iron & } 8	&c.	
Ox. Mang. }		

iii. Disseminated—amorphous.
iv. S. G. = 6.22.

Species 2.　Sulphuret of Zinc—*Blende.*

i. *Yellow.*

	Berg.	Hecht.
Zinc 64	44	62
Sulph. 20	17	21
Iron 5	5	3
Flu. Ac.... 4	....	....
Silica 1	24	....
Water 6	5	4
Alum.	5	2
Loss	....	2
Oxide of Zinc	44 5	60.4 } ?
Sulphur	16.4	22.5 }

Brown.

	Thom.	Theory?		Thom.
Zinc	58.8	Ox. Zinc 59.9+	1	58.64
Sulp.	23.5	Sulphur 22.3+	1	28.64
Iron	8.4	&c.		11.96
Silica	7.			.76
Loss	2.			

Black.

	Berg.	Theory.		Lamp.	Berg.
Zinc	52.	53.5+	1	53.	45
Sulp.	26.	24.4+	1	26.	29
Iron	8.	&c.		12.	9
Silica	6.			....	4
Water	4.			4.	6
Loss	4.			5.	6

II. *Prim.* Rhomboidal dodecahedron.

III. Crystals 50.—Concretionary—amorphous.

IV. S. G. = 4.1665.

IX. White vitriol.

Species 3. Carbonate of Zinc—*Calamine.*

	Smithson		Theory.		Smit.	Theory.	
I. Ox. Zinc..	65.2	64.8	66.1+	1	69	54	1
Car. Acid	34.8	35.2	33.8+	1	28	28	1
Iron	....	..	....		1	&c.	

II. *Prim.* Obtuse rhomboid.

III. Crystals — Earthy — pseudomorphous — amorphous.

IV. S. G. = 4.3.

IX. Usual ore for Zinc.

Species 4. Silicate of Zinc—*Electric Calamine.*

I.	Klap.	Theory.	Smiths.	Theory.	John.	Theory.	
Ox. Zinc 66	66.	68.	67.7+	69.25	69.25		1
Silica 33	24.5+	25.	25.2+	30.75	25.76+		1
... .	8.4+	..	&c.	....	4.98+		
Water....	....	4.4	....		....		
Loss........	....	2.3	....		...		

II. *Prim.* Rectangular octohedron.

III. Crystals 20. — Concretionary — stalactitic — amorphous.

IV. S. G. = 3.4.

GENUS XIV. ANTIMONY.

Printing types—optical mirrors.

Species 1. Native Antimony.

II. *Prim.* Regular octohedron.

III. Crystals (artificially formed) 3.—Disseminated —amorphous.

IV. S. G. = 6.7021 to 6.72.

VI. Argentiferous Antimony—silver 1 per cent.

Arsenical Antimony—arsenic 2 to 16 per cent.

Cupriferous Antimony.

	Theory.
Antimony ..70	70.
Copper 20	19.33⎱
Sulphur.... 9	9.66⎰
Arsenic.... 1	1

Nickeliferous Antimony.

	Vauq.
Antimony ...50	
Nickel 25	
Arsenic....	
Sulphur ...	25
Iron	
Lead	

Species 2. Oxide of Antimony.—*White Antimony.*

III. Tabular — acicular — capillary — earthy — amorphous.

VI. Ferro-plumbiferous.

Theory.

Antimony..34	34.	}	Ox. Antim..40.04 +
Oxygen ...16	18.61 +	}	Per-ox. Iron 34.57 +
Iron 22	22		Sulp. Lead .13.84 +
Lead 12	12	}	Silica ....14.
Sulphur ... 2	.84 +	}	102.45 +
Silica 14	14	}	
100	102.45 +		

IX. Colour for porcelain—pastes and enamels—medical properties of the two oxides.

Species 3. Sulphuret of Antimony.—*Grey Antimony.*

	Proust	J. Davy	Berg.	Theory.		
I. Antim..75.1	74.06	74	73.77 +	1	
Sulp....24.9	25.94	26	26.22 +	1	

II. *Prim.* Rhomboidal octohedron.

III. Crystals 20.—Acicular—capillary—amorphous.

IV. S. G. = 4.1327 to 4.5165.

VI. Argentiferous — auriferous — with crystals of sulphur.

With native Antimony?

Antimony .. 78.3 | 19.2 native.

Sulphur .... 19.7 | 59.1 } 19.7 } sulphuret. }

IX. Methods of extracting the metal—uses in the East.

Species 4. Sulphuretted Oxide of Antimony.— *Red Antimony—Kermes Mineral.*

	Klap.	Theory.	
I. Antimony ..	67.5	65.21 + }	1
Oxygen ...	10.8	23.18 + }	1
Sulphur ...	19.7	11.59 +	1

III. Acicular—amorphous.

IV. S. G. = 4.09.

Species 5? Bournonite.—*Endellione—Triple Sulphuret of Lead.*

	Hatchet		Theory.	
Antimony..	24.23	24.23	Sulp. Ant..	32.84 +
Lead	42.62	42.62	Sulp. Lead	49.17 +
Copper ...	12.8	12.8	Sulp. Cop..	16.
Iron	1.2	1.2	Sulp. Iron..	1.88 +
Sulphur ...	17.	19.04 +		99.89 +
	97.85	99.89 +	.: mean error = .40 +	

III. Crystals.

IV. S. G. = 5.7.

GENUS XV. SILVER.

S. G. $= 10.47$ to 10.51 — Leaf Silver $= \frac{1}{160000}$ inch thick.—Wire $\frac{1}{10}$ inch supports 370 lbs.—Alloy of commerce, Silver : Copper : : 137 : 7 and S.G. less than the mean by .1264—fulminating compounds—Arbor Dianæ—indelible ink.

Species 1. Native Silver.

II. *Prim.* Cube.

III. Crystals 3.—Granular—amorphous.

IV. S.G. $= 10$ to 10.338.

VI. Purest is alloyed with .03 of other metals.—

Auriferous—Silver. .36 to 72 ⎱
Gold . .64 to 28 ⎰ per cent.

Species 2. Sulphuret of Silver.—*Vitreous Silver.*

	Klaproth	Theory.		
I. Silver....75 to 85	77.46 +	1	
Sulphur..25 to 15	22.53 +	2	

II. *Prim.* Cube.

III. Crystals 3.—Amorphous.

IV. S.G. $= 6.9099$ to 7.215.

VI. Black silver—brittle sulphuretted silver—white silver—Bismuthic silver.

Arsenical sulphuretted Silver.

	Berg.	Theory.			Proust.
Sulphur ...	13	13.	2	Sulp. Silver ..	74.35
Silver	60	44.86+	1	Sulp. Ars.....	25.
....		15.13+		Sulphur and ⎫	
Arsenic ...	27	27.		Ox. Iron ...⎭	..65

IX. Ancient medals.

Species 3. Carbonate of Silver.

	Selb	Theory.	
I. Silver		4.5	.
Per-ox. Sil...	72.5	68. ⎫	1
Carb. Acid ..	12.	12. ⎭	1
Carb. Ant....	15.5	15.5	.

Species 4. Muriate of Silver—*Horn Silver.*

	Klap.	Theory.		Klap.	Theory.	
I. Silver ...	76.	27.24+		67.75	23.89 +	
....		48.75+	1	..	43.85+	1
Oxygen	7.6	7.09 +	2	6.75	6.37+	2
Mur. Acid	16.4	16.4	1	14.75	14.75	1
Ox. Iron	..	....		6.		
Sulp. Acid	..	....		...25	&c.	
Alumina .	..	....		1.75		

II. *Prim.* Cube ?

III. Crystal 1.—Acicular—investing—amorphous.

IV. S. G. = 4.745 to 4.8.

Species 5.   Antimonial Silver.

	Abich	Klap.	Theory.	
I. Silver ..	75.25	76 to 84	70.6 +	1
Antim...	24.75	24 to 16	29.3 +	1

III. Crystals—Plumose—granular—amorphous.

IV. S. G. = 9.4406 to 10.

Species 6.   Arsenical Silver?

Species 7.   Red Silver—*Ruby Silver.*—*Sulphuret of Antimonial Silver?*

	Lamp.	Klaproth			Theory.	
I. Silver..	61.	62.	60.	60	56.4 +	1
Oxygen	4.2	5.1	5.	4	4.1 +	1
Antim...	19.	18.5	20.3	19	23.0 +	1
Sulp. ..	13.3	14.4	14.7	17	16.4 +	2

	Then.	Vauq.	Lamp.
Silver..	58.	56.67	54.27
Oxygen ..		12.13	11.85
Antim...	23.5	16.13	16.13
Sulp. ..	16.	15.07	17.75
Loss...	2.5	..	..

II. *Prim.* Obtuse rhomboid P. P′. = 109° 28′.

III. Crystals — Dendritic—disseminated—amorphous.

IV. S. G. = 5.5637 to 5.5886.

GENUS XVI.   MERCURY.

S. G. of solid = 14.176 to 14.465—S. G. of fluid = 13.568.

Condensation = $\frac{1}{23}$ of its volume.

Impurities in the mercury of commerce—use in gilding—silvering—mirrors—barometer—thermometer.

Species 1.   Native Mercury—*Quicksilver.*

III. Liquid.

IV. S. G. = 13.581 to 13.6.

Species 2.   Sulphuret of Mercury—*Cinnabar.*

	Sage	Klap.		Theory.	
I. Mercury	..30	84.5	85.	86.20 +	1
Sulphur	..20	14.75	14.25	13.79 +	2

	Lamp.	Theory.	
Mercury	..81.	81.	1
Sulphur	..15.2	12.96	2
....	..	2.24	
Iron	..... 4.7	3.808	

II. *Prim.* Acute rhomboid.

III. Crystals 2. — Granular — pulverulent — amorphous.

IV. S. G. = 6.9 to 10.871.

VI. Hepatic Cinnabar—S. G. = 7.1 to 7.186.

	Klap.	Theory.	
Mercury	81.8	81.8	1
Sulphur	13.75	13.2	2
....		.55	
Copper	.02	&c.	
Ox. Iron	.2		
Alumina	.55		
Silica	.65		
Carbon	2.3		
Loss	.73		

Bituminous Cinnabar.

Species 3.   Muriate of Mercury—*Horn Quicksilver.*

	Klap.	Theory.	Klap.	Theory.	
I. Mercury	76.	29.9	67.75	5.7	
....		46.09⎱	..	61.2⎱	1
Mur. Acid ..	16.4	16.4 ⎰	21.	21. ⎰	2
Sulp. Acid ...	7.6		..	&c.	
Iron			6.		
Alumina			...5		
Lime			...25		
Loss			4.25		

	Berg.	Theory.	Berg.	Theory.	
Ox. Mercury ..	70	67.0+	75.	74.1+	1
Mur. Acid	20	23.9+	24.5	25.3+	2

III. Crystal 1.—Amorphous—tubercular crusts.

Species 4.   Argental Mercury—*Native Amalgam*
    —*Silver Amalgam.*

	Klap.	Theory.	Cordier	Theory.	
I. Silver 36	35.4+	27.5	27.5	1	
Mercury......64	64.5+	72.5	50.	1	

II. *Prim.* Regular octohedron?
III. Crystals 4.
IV. S. G. = 10. to 14.1192.

GENUS XVII. Bismuth.

S. G. = 9.8227—Easy method of obtaining it
crystallized—

Bism. : gold :: 1000 : 1
Bism. : lead : tin :: 8 : 5 : 3
Bism. : lead : tin : mercury :: 8 : 5 : 3 : 1 } Alloys.
Bism. : lead : tin : mercury :: 2 : 1 : 1 : 4

Pearl white — Mosaic silver — pewter — solder —
printing types.

Species 1.   Native Bismuth.

II. *Prim.* Regular octohedron.
III. Crystals 2—feathery—reticulated—amorphous.
IV. S. G. = 9.02 to 9.57.

Species 2.   Oxide of Bismuth—*Bismuth Ochre.*

III. Disseminated—amorphous.
IV. S. G. = 4.7.

Species 3.   Carbonate of Bismuth.

III. Amorphous.

Species 4.   Sulphuret of Bismuth.

I.   Lagerh. J. Davy.

Bismuth . . 81.25 | 81.7 | 1 ⎱
Sulphur . . 18.75 | 18.3 | 1 ⎰   Theory for the artificial.

        Sage.   Klap.   Theory.

Bismuth . . 60 | 60 | 68.9 + | 1 |
Sulphur . . 40 | 39 | 31.0 + | 2 |

III.   Acicular—capillary—amorphous.

IV.   S. G. = 6.131 to 6.4672.

VI.   Cupriferous.

	Klap.	Theory.	
Bismuth . . 47.24	47.24	Sulph. B. . . 57.89 +	
Copper . . 34.66	34.66	Sulph. C. . . 43.32 +	
Sulphur . . 12.58	19.31	&c.	
Loss 5.52	&c.		

    Or?

Bismuth . . . . 40.34 + |
Bisulph. Bis. . . . . 10.80 + |
Sulp. Cop . . . . . . 43.32 + |
&c.

Plumbo-cupreous.	Argentiferous *(needle ore.)*
John.	
Bismuth 43.20	Bismuth . . . . 27.
Lead 24.32	Sulphur 16.3
Copper 12.10	Silver . . . . 15.
Sulphur 11.58	Lead . . . . 33.
Gold 0.79	Copper 9
Nickel 1.58	Iron 4.3
Tellurium 1.32	Loss 3.5
Loss of Oxy- ⎱ genatedSulp. ⎰ 5.11	

GENUS XVIII. Cobalt.

S. G. = 8.5384—Alloy of Cobalt : Gold :: 1 : 65
—Zaffre and Smalt—imitation of sapphire,
lapis-lazuli, &c.—Sympathetic inks, blue, red,
and green.

Species 1. Per-oxide of Cobalt— *Earthy Cobalt.*

		Proust.	Theory.	
I.	Cobalt........	74	74.14	1
	Oxygen	26	25.86	2

III. Concretionary—pulverulent—amorphous.

IV. S. G. = 2. to 2.5—and 3.509?

Species 2. Bi-sulphuret of Cobalt—*Cobalt Pyrites.*

		Hising.		Theory.	
I.	Cobalt ..	43.2	43.2	Bisulp. Cob. ..73.33	
	Sulphur .	38.5	39.36	Bisulp. Cop. ..21.6	
	Copper .	14.4	14.4	Bisulp. Iron .. 5.56	
	Iron....	3.53	3.53	&c.	
	Silex ..	.33	.33		
		99.96	100.82		

Species 3. Sulphate of Cobalt—*Red Vitriol.*

	Koppe.	Theory.	
I. Ox. Cobalt..38.71	34.83+	1
Sulp. Acid ..19.74	23.61+	1
Water 41.55		

II. Investing—stalactitic.

Species 4. Arsenical Cobalt.

I. Arsenic........49.4+ | 1 } Theory.
   Cobalt ....50.5+ | 1 }

II. *Prim.* Cube.

III. Crystals—reticulated—concretionary—amorphous.

VI. *Var.* 1. Bright white Cobalt—*Cobalt glance.*

	Klap.	Theory.	Tassaërt	Theory.		
Arsenic 55.5	12.5+	49.	14.89+		
....	42.9+	....	34.10+	1
Cobalt ....44.	44.	36.66	36.66	1
Sulphur........ .5	&c.	6.5	&c.	
Iron	...	5.66		
Loss	....	2.18		

S. G. = 4.9411 to 6.466.

Var. 2. Grey Cobalt.

	Klap.	Theory.	
Arsenic........33	13.4+	
....	19.5+	1
Cobalt 20	20	1
Iron 24	&c.	
Bismuth } 23		
Earth, &c. }		

Var. 3.   Tin white Cobalt.

	John.	Laugier.	
Arsenic	65.75	50.	68.5
Cobalt	28.	12.7	9.6
Iron	5.	10.5	9.7
Mangan.	1.25	..	..
Sulphur	..	..	7.
Silex..........	..	25.	1.
Loss	..	..	4.2

S. G. = 7.379 to 7.751.

Species 5.   Arseniate of Cobalt—*Red Cobalt.*

	Buch.	Theory.	
i. Cobalt........	39	38.1 +	1
Ars. Acid	38	38.8 +	1
Water	23		

iii. Acicular—pulverulent—investing.

GENUS XIX. Gold.

S. G. = 19.2572 to 19.36—One grain may = 500 ft. in length, and can be extended over 56 square inches—Wire $\frac{1}{10}$ inch supports 500 lbs.—Standard, Gold : $\left. \begin{matrix} \text{silver} \\ \text{copper} \end{matrix} \right\}$ :: 11 : 1—Green gold, Gold : silver :: 4 : 1.

Gilding — colours for glass — purple powder of Cassius.

Species 1.   Native Gold.

II. *Prim.* Cube?

III. Crystals — capillary — ramose — pulverulent— amorphous.

IV. S. G. = 17. to 19.

VI. Argentiferous—Gold 64 to 93.5⎱ per cent.
                Silver 36 to 6. ⎰

IX. Amalgamation and Cupellation.

GENUS XX.   PLATINUM.

S. G. = 20.847 to 20.98 and 23.—Wire, diameter = $\frac{1}{5000}$ (and $\frac{1}{30000}$?) of an inch.

Species 1.   Native Platinum—*Platina.*

III. Crystal—In grains, some = 2 inches cubed.

IV. S. G. = 15.6017 to 19.5.

VI. Alloyed with Iron, Iridium, Osmium, &c. &c.

GENUS XXI.   PALLADIUM.

Species 1.   Native Palladium.

III. In grains.

IV. S. G. = 10.9 to 11.8.

VI. Alloyed with Platina, Gold, Iridium.

GENUS XXII. Tellurium.

S. G. = 6.11.

Species 1. Native Tellurium.

	Klaproth.	purest	graphic yellow	black	
i. Tellurium......	92.55	60	44.75	32.2	33.
Gold	.25	30	26.75	9.	8.5
Silver		11	8.5	0.5	0.5
Lead		...	19.5	54.	50.
Copper		....	....	1.3	0.5
Iron 7.2		...	....	....	...
Sulphur		..	0.5	3.	7.5
iv. Spec. Grav. 5.7 to 6.1	5.7	10.6	8.919		

ii. *Prim.* Rectangular prism.

iii. Crystals 20.

GENUS XXIII. Uranium.

S. G. = 6.44.

Species 1. Oxide of Uranium?—*Uran-ochre* —*Pitchblende.*

	Lampad.	Klap.	Sage.
i Ox. Uran......	32.	86.5	78.
Silica..........	56.	5.	..
Alumina	3.5	..	..
Iron..........	7.5	2.5	20.
Sulphur........	..	..	2.
Sulph. Lead	..	6.	..
Loss	1.	..	..

III. Globular—pulverulent—amorphous.
IV. S.G. = 7.575.

Species 2. Phosphate of Uranium—*Uranite*—*Uran-mica.*

I.		Gregor	R. Phill.	Theory.	
Ox. Uran.	.74.4	60.	Ox. Uran....60.		1
Phos. Acid.	..	16.	Phos. Acid....12.		1
Ox. Copper	8.2	9.	Water 7.7		2
Water	...15.4	14.5	Ox. Copper . 9.	1	
Loss	.... 2.	..	Phos. Acid.... 3.9 +	1	
			Water .... 2.5 +	2	
			Water 4.2		

II. *Prim.* Right symetrical prism.
III. Crystals—pulverulent.
IV. S.G. = 2.19 to 3.25.

GENUS XXIV. Titanium.

Species 1. Oxide of Titanium—*Titanite*—*Rutil.*

II. *Prim.* Right symetrical prism.
III. Crystals 2—acicular—fibrous *(hairs of Venus)*—pulverulent.
IV. S. G. = 4.18 to 4.24.
VI. Nigrine; S. G. = 4.4.

Ox. Titan............84.	
Ox. Iron ....14.	Klaproth.
Ox. Mang. .... 2.	

Menaccanite; S. G. = 4.4.

Ox. Titan............45.25
Ox. Iron ....51.
Ox. Mang........... .25
Silica 3.5
} Klaproth.

Iserine.

Ox. Titan....... ....48.
Ox. Iron ....48.
Ox. Uran............ 4.
} Thomson.

Species 2. Anatase; *Octohedrite.*

II. *Prim.* Symetrical octohedron P. P'. = 126° 47'.
III. Crystals 4.
IV. S. G. = 3.85.

Species 3. Sphene.

	Klap.	Cordier.	Klap.		Abild.
I. Ox. Titan..33.	33.3	45	74	58
Silica......35.	28.	36	8	22
Lime......33.	32.2	16	18	20
Water	..	1	..	..
Loss	6.5	..	..	..

II. *Prim.* Rhomboidal octohedron, P.P'. = 131° 16'.
III. Crystals—amorphous.
IV. S. G. = 3.23 to 4.24.

GENUS XXV. CERIUM.

Species 1. Silicate of Cerium?

Var. 1. Cerite.

I.	Hisinger.	John.	Theory?		Vauq.
Ox. Cerium..68.59	71.4	69.75	1	63.
Silica .....18.	18.	18.	1	17.5
Ox. Iron .... 2.	5.25	&c.		2.
Copper...... ..	.35			..
Lime 1.25	..			4.
Water..... 9.6	4.			12.

	Hisin.	Klap.	Thoms.
Ox. Cerium ..50.	54.5	44.
Silica........23.	34.5	47.3
Ox. Iron .....22.	3.5	4.
Lime........ 5.5	1.25	....
Water	5.	3.
Loss	....	1.7

III. Amorphous.

IV. S.G. = 4.53 to 4.93.

Var. 2. Allanite—*Cerin.*

	Thoms.	Berz.	Wollast.
I. Ox. Cerium ..33.9	28.19	19.8
Silica........35.4	30.17	34.
Ox. Copper	.87	..
Ox. Iron 25.4	20.72	32.
Alumina 4.1	11.31	9.
Lime 9.2	9.12	..
Water ...... 4.	..	..
Loss ....12.	..	..

III. Crystals?—amorphous.
IV. S. G. = 3.1 to 3.4.

Var. 3.   Orthite.

I.		
Ox. Cerium....17.39	19.44	
Silica 36.25	32.	
Ox. Iron 11.42	12.44	
Ox. Mang. 1.36	3.4	
Alumina 14.89	14.8	>Berzelius.
Lime 4.89	7.84	
Yttria 3.8	3.44	
Water 8.7	5.36	

III. Amorphous.
VI. Pyrorthite; Carbon = 25 per cent.

Species 2.   Fluate of Cerium?

Var. 1.   Sub-fluate of Cerium.
III. Amorphous.

Var. 2.   Neutral Fluate, or Deuto-fluate of
Cerium.

III. Six-sided prism—amorphous.

Var. 3.   Double Fluate of Cerium and Yttria.
III. Amorphous.

Var. 4.   Yttrocerite.

III. Amorphous.
IV. S. G. = 3.447.

GENUS XXVI.   COLUMBIUM—*TANTALIUM*.

Species 1.   Oxide of Columbium.—*Columbite —Tantalite*.

III. Crystals—amorphous.

IV. S. G. = 5.87 to 7.95.

Species 2.   Yttro-columbite—*Yttro-tantalite*.

	Berzelius.		
I. Ox. Columb....57.	60.12	51.81
Yttria 20.25	29.78	38.51
Ox. Iron 3.5	1.15	.55
Ox. Uran........ .5	6.62	1.11
Lime 6.25	.5	3.26
Tung. Acid.... 8.25	1.04	2.59

III. Crystals?—amorphous.

IV. S. G. = 5.13 to 5.88.

GENUS XXVII.   IRIDIUM.

Species 1.   Alloy of Iridium and Osmium.

III. Crystals 2.—In grains.

IV. S. G. = 19.5.

OSMIUM.

RHODIUM.

CADMIUM.

GENUS XXVIII. POTASSIUM.

Potash—Hydrate of Potash.

Species 1. Sulphate of Potash.

I.	Artificial.	Wollas.	Kirwan.	Berz.	Berard.	Berth.	Theory.
Hyd. Potash	54.2	54.8	55	57.24	58.5	58.76	1
Sul. Acid	45.8	45.2	45	42.76	41.5	41.23	1

II. *Prim.* Acute rhomboid P. P'. $= 87^\circ 48'$.

III. Crystals 2—amorphous.

Species 2. Nitrate of Potash.—*Nitre—Saltpetre.*

	Klap.	Theory.	
I. Hyd. Potash....	19.85	19.85	1
Nit. Acid	22.75	18.8	1
Lime	27.	&c.	
Carb. Acid	13.4		
Loss	1.4		

II. *Prim.* Rectangular octohedron P.P'. $= 111^\circ 14'$. M. M'. $= 120^\circ$.

III. Crystals 7—acicular—fibrous—incrusting.

GENUS XXIX. SODIUM.

Soda—Hydrate of Soda—Manufacture of Glass.

Species 1. Sulphate of Soda—*Glauber's Salt.*

I.	Artificial.	Woll.	Berard.	Theory.	
Hyd. Soda..	.44	47.22	49.3+	1	
Sulp. Acid..	.56	52.78	50.6+	1	

II. *Prim.* Symmetrical octohedron P. P′. = 100°.

III. Crystals 2—pulverulent—incrusting.

Species 2.   Bi-carbonate of Soda—*Natron*— *Trona.*

	Klap.	Theory.	
I. Hyd. Soda.........37.		36.17 +	1
Carb. Acid ....38.		38.82 +	2
Water 22.5		&c.	
Sulp. Soda..... 2.5			

II. *Prim.* rhomboidal octohedron P. P′. = 78° 28′?

III. Crystals?—fibrous—amorphous.

VIII. Recent formation in Stoke church, N. Devon.

Species 3.   Borate of Soda—*Borax*—*Tincal.*

	Klap.	Berg.	Theory.	
I. Hyd-Soda....14.5		17	16.77	1
Bor. Acid....37.		36	36.	4
Water ......47.		47	&c.	
Loss 1.5		..		

II. *Prim.* Oblique rectangular prism; P. M. = 106° 7′.

III. Crystals 5—amorphous.

Species 4.   Muriate of Soda—*Salt.*

I. Artificial.	Berard.	Berzelius.		Marc.	Kirw.	Theory.	
Hyd. Soda .. 57	54.26	53.44		54	53	52.56 +	1
Mur. Acid .. 43	45.74	46.55		46	47	47.43 +	1

Fossil.	Berg.	Theory.	Berg.	Theory.	
Hyd. Soda . .42	42.	50	40.1	1	
Mur. Acid . .52	34.5	33	33.	1	
Water. 6	..	17	..	.	

II. *Prim.* Cube.

III. Crystals—capillary—fibrous—amorphous.

IV. S. G. = 2.54.

VIII. Brine springs and salt lakes.

Species 5. Nitrate of Soda.

II. *Prim.* Obtuse rhomboid P. P'. = 106° 16'.

III. Amorphous.

IV. S. G. = 2.0964.

GENUS XXX. AMMONIUM?

Ammonia.

Species 1. Sulphate of Ammonia.

	Kirwan	Theory.		Phil.	Theory.	
I. Ammonia. .29.7	25.7+	1	40	36.9	2	
Sulp. Acid 55.7	60.6+	1	42	43.4	1	
Water 14.16	13.6	1	18	19.5	2	
Loss. 44	..		..			

III. Crystal 1—concretionary—pulverulent.

Species 2.   Muriate of Ammonia—*Sal Ammoniac.*

	Klap.		Theories.			
I. Ammonia ..	31.4	32.0	25.+	1	38.2	2
Mur. Acid ..	49.5	50.73	58.+	1	41.5	1
Water	16.6	17.	14.+	1	20.2	2
Sulphate of Ammonia	2.5	....				
Soda		.27				

II. *Prim.* Regular octohedron.

III. Crystals 3—plumose — concretionary — amorphous.

GENUS XXXI.   CALCIUM.

Lime—Hydrate of Lime.

  Species 1.   Carbonate of Lime—*Iceland Spar.*

I.	Berg.	Vauq.	Biot.	Holmes	Theory.	
Lime	55.	57.	56.351	55.9375	56.862 +	1
Carb. Acid	34.	43.	42.919	44.0625	43.137 +	1
Water ..	11.	..	.73	....	....	

II. *Prim.* Obtuse rhomboid $P'. P'. = 105° 5'.$

III. Crystals 300.

     Foliated (*Schiefer spar*)—fibrous (*Satin-spar.*)

     Granular limestone (*statuary marbles*)—Parian —Thasian—Naxian—Pentelican — Carrara —Cipolino.

     Earthy carbonate of lime—incrustations ; mode of deposition.

Stalactites—Theory of their formation—Alabaster.

Compact limestone (*secondary marbles*)—Lumacella —— Cotham —— Carinthia —— Ruin marble—Madreporite —Ludus Helmontii—Swinestone—Bituminous limestone—Argillo-ferruginous limestone (*Calp*) — Pisolite — Oolite—Chalk—Agaric mineral—Aphrite—Tufa—Marl.

iv. S. G. of crystals = 2.69 to 2.7.

S. G. of the compact varieties = 2.3 to 2.8.

vi. Fontainbleau spar ; S. G. = 3.6.

	Sage.	Theory.	
Lime......18.5	18.7 +	I
Carb. Acid .14.5	14.2 +	I
Sand......67.	....	

∴ Atoms of Silica : Atoms of Carb. Lime, in a greater ratio than 6 : 1.

ix. Architecture — statuary — mortar — agriculture —lithography—drawing chalks—whiting—casts made at St. Philip in Tuscany.

Species 2. Arragonite—*Hard calcareous spar.*

	Fourc.	Biot.	Holmes	Theory.	
i. Lime.....58.5	56.327	55.8	56.862 +	I
Carb. Acid 41.5	43.045	43.4	43.137 +	I
Water		0.628	0.8	....

ii. *Prim.* Rectangular octohedron P.P′. = 109° 28′.

III. Crystals — Fibrous — coralloidal *(flos-ferri)*— stalactitic—amorphous.

IV. S. G. = 2.9267.

Species 3.   Sulphate of Lime—*Selenite.*

	War.	Bucholtz.		Phil.	Bert.	Theory.	
I. Lime 32	33.9	33.		32.7	32.8	33.3	1
Sulp. Acid 47	43.9	43.5		46.3	45.2	45.9	1
Water ..21	21.	21.		21.	22.	20.6	2
Loss	1.2	2.5		..	..	..	

II. *Prim.* Right irregular quadrangular prism.

III. Crystals 10—Lenticular—acicular—fibrous— granular (*gypsous alabaster*)—compact (*gypsum*)—earthy.

IV. S. G. = 2.2642 to 2.3117.

VI. Calciferous ; carbonate lime = .012 per cent.

IX. Statuary—vases, &c.—ancient lacrymatories— Temple of Seia—Plaster of Paris—agriculture.

Species 4.   Anhydrous Sulphate of Lime— *Anhydrite—Muriacite.*

	Vauq.	Klaproth.		Theory.	
I. Lime......40.	42.	42.		42.0+	1
Sul. Acid ..60.	56.5	57.		57.9+	1
Silex	..	.25		....	
Soda	.25	..		....	
Ox. Iron	..	1.		....	

II. *Prim.* Right rectangular prism.

III. Crystals 3—fibrous—amorphous.

IV. S.G. = 2.5 to 2.9.

VI. Siliciferous (*vulpinite*); Silica = 8 per cent.

IX. Used as a marble.

Species 5. Nitrate of Lime.

	Berg.	Kirwan.	Theory.		
I. Lime	32	32.	31.4+	1	
Nit. Acid	43	57.44	58.7+	1	Artificial.
Water	25	10.56	9.8+	1	

III. Crystals—acicular.

Species 6. Phosphate of Lime—*Apatite—Asparagus stone.*

	Klap.		Vauq.	Theory.		
I. Lime	55	53.75	54.28	50.8+	1	
Phos. Acid	45	46.25	45.72	49.1+	1	

II. *Prim.* Regular hexahedral prism.

III. Crystals—granular—compact (*phosphorite*)—pulverulent.

IV. S. G. = 3.09 to 3.2.

VI. Quartziferous—Calciferous.

IX. Building material in Spain.

Species 7. Fluate of Lime—*Fluor spar.*

I.	Klap.	Thoms.	Rich.	Theory.	
Lime	67.75	67.34	65	63.0+	1
Flu. Acid	32.25	32.66	35	36.9+	1

F

II. *Prim.* Regular octohedron.

III. Crystals 70—most complex with 234 planes— nodular *(Blue John)*— amorphous *(Chlorophane)*—earthy.

IV. S. G. = 3.09 to 3.19.

VI. Quartziferous—aluminiferous.

IX. Flux—artificial gems—Vases, &c.— etching upon glass.

Species 8.   Bi-silicate of Lime—*Tabular spar —Wollastonite.*

I.	Seybert	Theory.	Klap.	Theory.	Bronsdorff.
Silica	51.	50.88 +	50.	49.83 + 2	52.58
Lime	46.	46.11 +	45.	45.16 + 1	44.45
Water	1.	....	5.	....	..
Alum. and Ox. Iron . }	1.33	....	..	....	.. 1.13
Magnesia and Loss }	.67	....	..	....	.68 1.16

II. *Prim.* Rectangular octohedron P.P′. = 139° 42′. M. M′. = 92° 18.

III. Crystals—amorphous.

IV. S. G. = 2.86.

Species 9.   Siliceous Borate of Lime—*Datholite.*

I.	Vauq.	Theory.	Klap.	Theory.	
Lime	34.	28.565	35.5	31.6	1
Silica	37.66	31.5	36.5	34.8	1
Bor. Acid............	21.67	21.67	24.	24.	1
Water	5.5	....	4.	&c.	

II. *Prim.* Right rhomboidal prism; M. T. = 109° 28'.

III. Crystals 3—Concretionary *(Botryolite)*—amorphous.

IV. S. G. = 2.98.

Species 10.   Arseniate of Lime—*Pharmacolite.*

I.	Klap.	Theory.	
Lime ....25.	25.17+	1
Ars. Acid 50.54	50.36+	1
Water 24.46		?

III. Acicular—capillary—mamillary.

IV. S. G. = 2.54 to 2.6.

Species 11.   Tungstate of Lime—*Tungsten.*

I.	Scheele.	Klaproth.		Berzel. and Theory.		
Lime......... ..31.	18.7	17.6	19.460	19.46+	1
Tung. Acid 65.	72.25	77.75	80.417	80.53+	1
Iron	1.25	....			
Mangan.	.75	....			
Silica .... 4.	1.5	3.			

II. *Prim.* Acute octohedron P. P'. = 130° 20'.

III. Crystals 3—amorphous.

IV. S. G. = 5.5 to 6.

Species 12.   Anhydrous Sulphate of Soda and Lime—*Glauberite.*

I.	Brog.	Theory.	
Anhyd. Sulp. Soda............51.	51.06+	1
Anhyd. Sulp. Lime ....49.	48.93+	1

ii. *Prim.* Oblique rhomboidal prism.
   P. M. = 104° 30′;   M. T. = 80° 8′.

iii. Crystals 3.

iv. S. G. = 2.73.

GENUS XXXII. Magnesium.

Magnesia.

Species 1. Hydrate of Magnesia.

I.

	Bruce.	Theory.	Vauq.	Theory.	
Magnesia	70.	68.9 +	64.	64.	1
Water	30.	31.0 +	29.	28.8	1
Silica		&c.	2.	&c.	
Iron			2.5		
Loss			2.5		

ii. *Prim.* Right symmetrical prism?

iii. Amorphous.

iv. S. G. = 2.13.

vi. Siliciferous (*Meerschaum*); Silica from 5. to 54. per cent.

ix. Keffekil pipes.

Species 2. Sulphate of Magnesia—*Epsom Salt.*

I.

	Berz.	Theory.	Klap.	Theory.	
Magnesia	19	16.5	33	33.33 +	1
Sulp. Acid....	33	33.	67	66.66 +	1
Water of Crystallis. }	48				

II. *Prim.* Right symmetrical prism.
III. Crystals 4—fibrous—granular—pulverulent.

Species 3.   Carbonate of Magnesia—*Magnesite.*

I.	Mitch.	Buch.	Theory.	Buch.	Theory.	
Magnesia ...47.5	46.59	46.45	48	47.2	1	
Carb. Acid 51.	51.	51.	52	52.	1	
Silica	.16	&c.				
Alumina	1.					
Iron	.25					
Water 1.5	1.					

	Bert.	Theory.		Bert.	Theory.		Bert.
Magnesia 23.	23. }		1	35.	34.	1	25.5
Car. Acid 36.	25.3 }		1	37.4	37.4	1	10.5
....	10.6 } 1.			....	&c.		....
Lime ....14.	14. } 1.			..			....
Water 4.5	&c.			1.			12.
Silica 20.6				26.			52.

	Bert.	Theory.		Klap.	Theory.	
Water 4.8	2.1			3	1.44	
... .	2.7 } 1.			..	1.55 } 1.	
Magnesia ....44.	6. } 1.			48	3.45 } 1.	
....	38. }	1		..	44.54 }	1
Carb. Acid 41.8	41.8 }	1		49	49. }	1
Silica 9.4	&c.			..	..	

III. Sub-granular — spongiform — pulverulent — amorphous.
IV. S. G. = 2.1751.

Species 4. Carbonate of Lime and Magnesia—Bitterspar—Pearlspar.

I.

	Hising.	Theory.				Klap.	Theory.		
Lime........	29.8	29.8 } 1.	1		Lime........	29	29 } 1	1
Carb.Acid	47.6	22.6 } 1.			Carb.Acid	23	22 } 1	
....		23.7 } 1.	1		Carb.Mag.	45	45	1
Magnesia	21.6	21.6 } 1.	1		Iron	3	&c.	
Iron	1.	&c.						

II. *Prim.* Obtuse rhomboid P. P′. $= 106^0$ 15′.

III. Crystals—amorphous.

IV. S. G. $= 2.48$ to 2.88.

VI. Miemite; S. G. $= 2.8$.

	Klap.	Theory.		
Carb. Lime..............	53.	52.2	1
Carb. Magnesia	42.5	43.2	1
Carb. Iron and Mang. ..	3.		

Dolomite—Conite—Magnesian Limestone (*compact Dolomite*).

Species 5. Borate of Magnesia—*Boracĭte.*

I.

	Pfuff.	Theory.		Westr.
Magnesia	30.68	42.62	1	13.5
Bor. Acid	54.54	54.54	1	68.
Silica	2.27	&c.		2.
Alumina				1.
Lime				11.
Iron	.57			.75
Loss	11.94			3.75

II. *Prim.* Cube.

III. Crystals 5.

IV. S. G. $= 2.56$.

Species 6. Silicate of Magnesia—*Chrysolite—Peridot.*

I.

	Vauq.	Theory.		Klap.	Chen.
Silica	38.	39.33	1	39.	39.
Magn.	50.5	49.16	1	43.5	53.
Iron	9.5	&c.		19.	7.5
Loss	2.			..	

II. *Prim.* Right rectangular prism.

III. Crystals 6—granular.

IV. S. G. = 3.4.

VI. Olivine; S. G. = 3.26.

	Klap.	Theory.		Klaproth.			Howard.	Theory.	
Silica..	41.	10.2		50.	52.	54	32.4		
....		30.8 } 1		..	..	..	21.6 }	1	
Magn.	38.5	38.5 } 1		38.5	37.75	27	27. }	1	
Iron ..	18.5	&c.		12.	10.75	17	&c.		
Lime ..	..			.25	.12		1 Nickel		
	meteoric				decomp.		1 Loss		

GENUS XXXIII. BARIUM.

Barytes.

Species 1. Sulphate of Barytes—*Heavy Spar.*

I.

	With.	Klap.	Bert.	Strom.	Theory.		
Barytes....	67.2	67	66	65.8	66.10+		1
Sulp. Acid	32.8	33	34	33.87	33.89+		1
Iron		..	..	.05			
Water		..	..	.05			
Loss		..	..	.32			

ii. *Prim.* Right rhomboidal prism, M. T. = 101°
42′.

iii. Crystals 73 — fibrous — radiated *(Bolognian
stone)* — concretionary — compact *(Cawk);*
S. G. = 4.8—Earthy.

iv. S. G. = 4.3 to 4.7.

vi. Fetid *(Hepatite.)*

ix. Bolognian phosphorus *(pyrophorus)* — white
paint.

Species 2.   Carbonate of Barytes—*Witherite.*

I.   Pell. Vauq. Klap. & Theory.

	Pell.	Vauq.	Klap.	& Theory.
Barytes........	62	74.5	78	1
Carb. Acid	22	22.5	22	1
Water	16	..	..	

ii. *Prim.* Obtuse rhomboid P. P′. = 93° 54′.

iii. Crystals 5—acicular —sub-fibrous — stalactitic
—amorphous.

iv. S. G. = 4.2919 to 4.3.

ix. Poisonous qualities.

GENUS XXXIV. Strontium.
Strontian.

Species 1.   Sulphate of Strontian—*Celestine.*

I.   Vauq. Klap. Stromyer. Theory.

	Vauq.	Klap.	Stromyer.	Theory.		
Strontian..	54	58	56.39	56.5+	1	
Sulp. Acid	46	42	42.95	43.4+	1	

II. *Prim.* Right rhomboidal prism, M. T. = 104° 48'.

III. Crystals 3 — acicular — fibrous — stellated — amorphous—pseudomorphous.

IV. S. G. = 3.4 to 4.

VI. Calcariferous; Carb. Lime = 8 per cent.

Species 2. Carbonate of Strontian—*Strontianite.*

I.	Klap.	Hope.	Pell.	Stromyer.	Theory.	
Strontian ...69.	61.21	62	70.313	70.27	1	
Carb. Acid 30.	30.2	30	29.687	29.72	1	
Water 5	8.59	8	....	....	.	

II. *Prim.* Obtuse rhomboid P. P'. = 99° 35'.

III. Crystals 3—fibrous—stellated—amorphous.

IV. S. G. = 3.65.

GENUS XXXV. ALUMINUM.

Species 1. Alumina—*Corundum—Sapphire.*

II. *Prim.* Acute rhomboid, P. P'. = 86° 4'.

III. Crystals 13—Laminar—fusiform.

IV. S. G. = 3.975 to 4.3.

VI. Adamantine spar (*imperfect corundum*); Alumina = 86 per cent.

Emery.

	Tenn.	Vauq.
Alumina	50	70
Ox. Iron........	32	30
Silex	8	
Residue ....,..	4	

IX. Jewelry—polishing powder.

Species 2.   Hydrate of Alumina?—*Diaspore.*

I.

	Vauq.	Theory.	
Alumina..........	80	51	1
Water	17	17	1
Iron	3		

II. *Prim.* Right rhomboidal prism, M. T. = 130°.
III. In laminated masses.
IV. S. G. = 3.4324.
VI? Gibbsite; stalactitic; S. G. = 2.4.

	Torrey.	Theory.	
Alumina........	64.8	52.05	1
Water	34.7	34.7	1
Loss	.5		

Species 3.   Sulphate of Alumina—*Alum.*

I.

	Klap.
Alumina	15.25
Sulp. Acid and Water ....}	77.
Potash	.25
Ox. Iron	7.5

ii. *Prim.* Regular octohedron.

iii. Crystals 5—capillary—fibrous (*plumose alum*)
—concretionary—amorphous.

ix. Uses in dying.

Species 4. Subsulphate of Alumina?— *Aluminite—Websterite.*

I.

	Strom.	Simon.	Buch.
Alumina	30	32.5	31.
Sulp. Acid	24	19.25	21.5
Water	45	47.	45.
Silex...........		.43	1.
Carb. Lime		.35	.5
Iron		.45	.5
Loss		....	.5

iii. Mamillary—pulverulent—amorphous.

iv. S. G. = 1.6.

vi. Siliciferous subsulphate of Alumina?

Water ....88.1 ⎫
Alumina .... 6.5 ⎬ Dr. Henry.
Sulp. Acid .... 3. ⎪
Silica .... 2.4 ⎭

Species 5. Alkilino-subsulphate of Alumina?

Alumina............38.654 ⎫
Sulp. Acid ....35.495 ⎬ Cordier.
Potash ....10.021 ⎪
Water & Loss 14.830 ⎭

ii. *Prim.* Acute rhomboid?

iii. Crystals 2.

iv. S. G. = 2.7517.

Species 6.　Phosphate of Alumina—*Wavellite*
—*Hydrargillite.*

	Berzel.	Theory.		Klapr.		Davy.
I. Alumina	35.35	33.30	1	71.5	68.	70.
Phos. Acid ..	33.40	33.40	1	....	....	...
Water	26.80	21.47	2	28.	26.5	26.5
Flu. Acid....	2.06	&c.		....	....	....
Lime ...,...	.50			....	...	1.4
Ox. Iron and⎱ Ox. Mangan.⎰	1.25			5	1.	,...

II. *Prim.* Oblique rhomboidal prism?

III. Crystals 2. — Acicular — fibrous — radiated globules.

IV. S. G. = 2.2 to 2.7.

VI. Crystalline character of Wavellite, impressed on clay-slate.

Species 7.　Silicate of Alumina?　*Cyanite*—
Disthene—Sappare.

	Sauss.	Theory.	Laugier.	Theory.	
I. Alumina ..	54.5	53.58	55.5	55.5	1
Silica	30.62	31.54	38.5	31.9	1
Lime	2.02	&c.	.5	&c.	
Magnesia..	2.3		...		
Ox. Iron ..	6.		2.75		
Water and⎱ Loss ....⎰	4.56		2.75		

II. *Prim.* Oblique irregular four-sided prism,
P. M. = 106° 55′; M. T. = 106° 15′.

III. Crystals 5—Acicular.
IV. S. G. = 3.517.
VI. Rhætizite.

APPENDIX ?

Fibrolite.

I. Alumina	.58.25	46	
Silica	...38.	33	
Ox. Iron	.75	13	Chenevix.
Loss	.... 3.	8	

III. Fibrous.
IV. S. G. = 3.2.

Species 8.   Fluate of Soda and Alumina—
   Cryolite.

	Klap.	Vauq.	Theory.		
I. Alumina	24	21	24.6 }1	1	
Flu. Acid	40	46	15.5 }1		
....	..	..	31.1 }2	1	
Soda	36	33	29.3 }1		
	100	100	100.5		

II. *Prim.* Right rectangular four-sided prism ?
III. Fibrous—amorphous.
IV. S. G. = 2.949.

Species 9.   Mellate of Alumina—*Mellite*

I. Alumina	 16	
Mellitic Acid	}84	Klaproth.
and Water		

II. *Prim.* Symmetrical octohedron, P. P'. = 93° 2'.
III. Crystals 3.—Granular.
IV. S. G. = 1.6.

GENUS XXXVI. Zirconium.

Zirconia.

Species 1. Silicate of Zirconia—*Zircon.*

I. *Jargoon.*

	Klap.	Theory.	Klapr.	
Zirconia ..69.	69.	1	68.	64.5
Silica 26.5	24.5	1	31.5	32.5
....	2.		..	....
Ox. Iron .. .5	&c.		...5	1.5
Loss ..... 4.			..	....

Hyacinth.

	Klap.	Theory.	Vauquelin.		
Zirconia 70.5	70.	1	66	65.5	64.4
Silica ....25.	24.88	1	31	31.	32.
....	.11		....	....	....
Ox. Iron 5	&c.		2	1.5	2.
Loss............ 4.5			1	2.	2.

Zirconite.

	Klap.	Theory.	John.	Theory.	
Zirconia ...65.	65.	1	64.	64.	1
Silica ......33.	23.1+	1	34.	22.7+	1
....	9.8+		....	11.2+	
Ox. Iron... ..	&c.		.25	&c.	
Magnesia .. 1.			....		
Ox. Titanium..			1.		

II. *Prim.* Symmetrical octohedron, P. P'. $= 84^\circ 20'$.

III. Crystals 45.—Granular.

IV. S. G. $= 4.38$ to 4.41.

GENUS XXXVII. YTTRIUM.

Species 1. Silicate of Yttria?—*Gadolinite.*

	Berzelius			Vauq.	Klap.	Ekeberg.
Silica	29.18	24.16	25.8	25.5	21.25	23.
Yttria ...	47.3	45.93	45.	35.	59.75	55.5
Ox.Cerium	3.4	16.9	16.69	....	....	....
Ox. Iron..	8.	11.34	10.26	25.	17.5	16.5
Ox. Mang.	1.3	..	....	2.	....	....
Lime	3.15	....	....	2.	Alu. 5	....
Glucina ..	2.	....	....	....	....	....
Water ...	5.2	....	....	....	.5	....
Loss	&c.	.6	.6	10.6	....	.5

Theory.				
Silica	18.92	18.37	18.	1
Yttria ...	47.3	45.93	45.	1

II. *Prim.* Oblique rhomboidal prism, M.T. = 109° 28'.

III. Crystal 1.—Amorphous.

IV. S.G. = 4.

GENUS XXXVIII. GLUCINUM.

Glucina.

Species 1. Emerald—*Beryl.*

	Vauq.	Theory.	Vauq.	Theory.	
Silica	64.6	64.7+	68.	68.5+	8
Alumina	14.	13.6+	15.	14.4+	1
Glucina	13.	13.2+	14.	14.0+	1
Lime	2.56	&c.	2.	&c.	
Ox. Iron	....		1.		
Ox. Chrome....	3.5		...		

	Berz.	Theory.		Klap.		Gmelin
Silica	68.35	60.	8	68.5	66.45	54.75
Alumina ...	17.6	14.85	1	15.75	16.75	24.41
Glucina	13.13	13.13	1	12.5	15.5	15.4
Iron	.72	&c.		1.	.6	1.5
Tantal.	.27			....	....	&c.
Ox. Chrome	....			3.	....	
Lime	....			.25	....	

II. *Prim.* Regular hexahedral prism.

III. Crystals 8.—Cylindrical.

IV. S. G. = 2.7227 to 2.7755.

Species 2.　Euclase.

	Vauq.	Berzelius.
I. Silica	35.	43.22
Alumina	22.	30.55
Glucina	12.	21.78
Ox. Iron	3.	2.22
Ox. Tin		.7
Loss	28.	....

II. *Prim.* Oblique rectangular prism, P. M. = 130° 8′.

III. Crystals 2.

IV. S. G. = 3.0625.

MINERALS

WHOSE GENUS IS UNCERTAIN.

———◆———

Species. Topaz.

	Klaproth.		Berz.	Vauquelin.		
i. Silica	35.	44.5	34.01	28.	39.	31.
Alumina ..	59.	47.5	58.38	47.	49.	68.
Flu. Acid..	5.	7.	7.75	17.	20.	..
Iron......	..	.5	..	4.	..	..
Loss......	1.	.5	..	4.	2.	1.

ii. *Prim.* Rectangular octohedron, P. P′. = 88° 2′, M. M′. = 122° 42′.

iii. Crystals 37—amorphous.

iv. S. G. = 3.5311 to 3.564.

vi. Pyrophysalite—*Physalite;* S. G. = 3.41.

	Hisinger	Berzel.
Silica	32.88	34.36
Alumina ..	53.35	57.74
Flu. Acid ..	..	7.77
Lime	.88	..
Iron	.88	..
Loss	12.11	..

Pycnite.—*Schorlaceous Beryl;* S.G. =3.48.

	Buch.	Berz.	Klap.	Vauq.	
Silica ....34.	38.43	43.	30.	36.
Alumina ..48.	51.	49.5	60.	52.6
Flu. Acid ..17.	8.84	4.	6.	5.8
Iron		..	1.	..	..
Water		..	1.	1.	..
Lime			..	2.	3.3
Loss			1.5	1.	1.5

Species. Spinelle—*Spinelle Ruby.*

	Vauq.	Klap.
I. Alumina 82.47		74.5
Silica		15.5
Magnesia.... 8.75		8.25
Lime		.75
Ox. Iron		1.5
Chr. Acid.... 6.18		..
Loss 2.57		..

II. *Prim.* Regular octohedron.

III. Crystals—granular.

IV. S.G. = 3.645 to 3.76.

VI. Pleonaste—*Ceylanite—Blue Spinelle.*

	Desc.	Berz.
Alumina ....68.		72.25
Silica ...... 2.		5.48
Magnesia ....12.		14.63
Ox. Iron ....16.		4.26
Loss .... 2.		..

Zinciferous Spinelle—*Automalite*—*Gahnite;*
S.G. = 4.7.

	Ekeberg.	Vauq.
Alumina	60.	42.
Silica	4.75	4.
Ox. Iron	9.25	5.
Ox. Zinc	24.15	28.
Loss	1.75	..

Species.   Chrysoberyl—*Cymophane.*

	Klapr.	Achard.
I. Alumina	75.5	64.
Silica	18.	15.
Lime	6.	17.
Ox. Iron	1.5	1.
Loss	3.	

II. *Prim.* Right rectangular four-sided prism.

III. Crystals 5—granular.

IV. S.G. = 3.796.

Species.   Tourmaline—*Schorl.*

	Vauquelin.			Hern.	
Silica	42.	45.	47.27	47	
Alumina	40.	30.	45.46	28	
Soda	10.	10.	....	..	*Rubellite*
Lime	..	..	1.78	7	*Siberite.*
Magnesia	..	..	....	10	
Ox. Mang.	7.	12.	5.49	2	
Loss	1.	2.	....	6	

	Vauq.	Berg.	*Green Tourmaline*	Klap.		Buch.	*Common Tourmaline*
Silica	40.	37.		36.75	36.5	35.5	
Alumina	39.	39.		34.5	33.75	33.25	
Potash	....	....		6.	...	....	
Lime	3.84	15.		....	.25	.5	
Magnesia	....	....		.25	6.08	9.3	
Ox. Iron	12.5	9.		21.	8.	5.10	
Ox. Mang.	2.	....		....	...	....	
Water	....	....		....	1.5	....	
Loss	2.66	....		....	13.92	16.35	

II. *Prim.* Obtuse rhomboid, P. P′. = 133° 26′.
III. Crystals 18—acicular—capillary.
IV. S. G. = 2.87 to 3.4.
VI. Indicolite.

Species. Garnet.

		Klaproth		Vauq.	
Silica	35.75	40.	43.	36.	*Precious.*
Alumina	27.25	28.5	15.5	22.	*Pyrope.*
Ox. Iron	36.	16.5	29.5	41.	*Almandine*
Lime	....	3.5	1.75	3.	*Syrian.*
Ox. Mang.	.25	.25	.5		
Magnesia	....	10.	8.5		

		Klapr.	Vauq.	Hesinger.	Weig.	
Silica	35.5	44.	34.	34.53	36.5	*Melanite.*
Lime	32.5	33.5	33.	34.26	30.8	
Ox. Iron	24.25	12.	25.5	36.05	28.7	
Alumina	6.	8.5	6.4	1.	....	
Ox. Mang.	.4	....	....	....	...	
Water	....	....	....	.5	4.	
Loss	....	2.	1.1	3.56	....	

	Laugier.	Simon.	Vauq.	Rose.	Vauq.	John.
Silica	..40.	35.	35.	37.	52.	35.2
Alumina	.20.	15.	8.	5.	20.	.2
Lime	...14.5	29.	30.5	30.	7.7	24.7
Ox .Iron	.14.	7.5	17.	18.5	17.	26.
Ox.Mang.	2.	4.75	3.5	6.25	....	8.6
Magnesia	....	6.5	....	....	....	....
Water	.. ....	1. }	....	....	....	1.05 } Soda
Loss	.... 5.	.75	....	....	3.3	2.25
	Aplome	Colophonite	Allochroïte		Common	

Pyrenite—Grossular—Topazolite.

II. *Prim.* Rhomboidal dodecahedron.

III. Crystals—granular—amorphous.

IV. S. G. = 3.56 to 4.19.

VI. Manganesian garnet; S. G. = 3.6—Ox. Mang. = 35 per cent.

Species. Cinnamon Stone—*Essonite.*

	Klap.	Lamp.
I. Silica 38.8		42.8
Alumina 21.2		8.6
Lime 31.25		3.8
Zirconia		28.8
Ox. Iron 6.5		3.
Potash		6.
Water		2.6
Loss 2.25		4.4

II. *Prim.* Right rhomboidal prism, M. T. = 102° 40′.

III. Amorphous fragments.

IV. S.G. = 3.6.

Species. Idocrase—*Vesuvian.*

	Klap.	Klap.
I. Silica	35.5	42.
Alumina	22.25	16.25
Lime	33.	34.
Ox. Iron	7.5	5.5
Ox. Mang	.25	....
Loss	1.5	2.25

Or,

	Theory.	
Sil. Alum.... 35.3 +	35.3 +	1
Sil. Lime ....51.2+	36.9+	1
....	14.2+	
Silica .... 4.1 +	&c.	
&c.		

II. *Prim.* Right symmetrical prism.

III. Crystals 9—amorphous.

IV. S. G. $= 3.088$ to 3.409.

VI. Gehlenite; S.G. $= 2.98$.

Silica 29.64	
Lime 35.5	
Alumina ..24.8	>Fuchs.
Ox. Iron .. 6.56	
Loss 3.3	

Species.   Staurolite—*Staurotide—Grenatite.*

	Klap.		Vauq.	
I. Alumina 52.25	41.	44.	47.06
Silica ....27.	37.5	33.	30.59
Ox. Iron ....18.	18.25	13.	15.3
Ox. Mang........ .25	.5	1.	....
Lime	...	3.84	3.
Loss .... 2.	2.75	5.16	4.05

II. *Prim.* Right rhomboidal prism, M. T. =
129° 20′.

III. Crystals 5.

IV. S. G. = 3.20.

Species.   Diopside—*Mussite—Alalite.*

	Laugier.	Theory.	*Or,*	
Silica ....57.	51.9 +	4	Bisil. Lime . 34.7 +	1
Lime 16.5	23.5 +	1	Bisil. Magn. 29.5 +	1
Magnesia ...18.25	16.2 +	1	.... 17.9 +	
Ox. Iron & } 6.	&c.		&c.	
Ox. Mang. }				
Loss 2.25				

II. *Prim.* Oblique rhomboidal prism, M. T. =
92° 55′.

III. Crystals.

IV. S. G. = 3.31.

VI. Sahlite.

	Vauq.	Theory	Hising.	Theory.	
Silica	53.	52.1+	54.18	53.6+	4
Lime	.20.	23.6+	22.72	24.3+	1
Magnesia	19.	16.2+	17.81	16.8+	1
Alumina	3.	&c.	...	&c.	
Ox. Mang.	4.		1.45		
Ox. Iron	...		2.18		
Loss	1.		1.2		

Augite—*Pyroxene.*; S. G. $= 3.226$ to 3.6.

	Tromsd.	Klap.	Simon.	Seybert.	Roux.
Silica	54.	55.	52.	50.33	45.
Lime	16.2	12.5	25.5	19.33	30.5
Magnesia	14.	13.75	7.	6.83	....
Alumina	3.05	5.5	3.5	1.53	3.
Ox. Iron	7.	11.	10.5	20.4	16.
Ox. Mang.	2.	....	2.25	....	5.
Potash	5.18	....	....	....	....
Water	....	1.	.5	.66	....
Loss	....	....	....	.92	.5

Coccolite.

	Vauq.		Theory.	
Silica	50.	Bisil. Mag.	26.	1
Lime	24.	Bisil. Lime	30.5	1
Magnesia	10.		19.9	
Alumina	1.5	Silica	13.5	
Ox. Iron	7.	&c.	....	
Ox. Mang.	3.			
Loss	4.5			

Species.   Hornblende—*Amphibole.*

I.	Laugier.	Klap.	Berg.
Silica ....42.		47.	58.
Lime.............. 9.8		8.	4.
Magnesia 10.9		2.	1.
Alumina 7.69		26.	27.
Ox. Iron 22.69		15.	9.
Ox. Mang........ 1.15		....	....
Water 1.92		....	....
Loss 3.85		2.	1.

II. *Prim.* Oblique rhomboidal prism, M. T. = 124° 36′.

III. Crystals 11 — acicular — fibrous—granular—amorphous.

IV. S. G. = 2.5 to 3.6.

VI. Pargasite ; S. G. = 3.11.

Silica 42.01	
Alumina 14.08	
Lime............ ..14.28	
Magnesia.... ..18.27	Bonsdorff.
Ox. Iron 3.52	
Ox. Mang... .... 1.02	
Water 3.9	
Loss 3.92	

Actynolite — *Actinote* — *Strahlstein*; S. G. = 3.33.

		Berg.	
Silica ....50.	72.	54.
Alumina .... .75	2.	27.
Lime. 9.75	6.	.33
Magnesia 19.25	12.	20.
Ox. Iron ....11.	7.	4.
Ox. Chrome. .. 5.		
Water .... 3.		
Loss 1.25		

Tremolite; S. G. = 2.92 to 3.2.

	Chen.	Klap.	Lowitz.		Lowitz.	
Silica ....27.	65.	52.		44	
Lime ....21.	18.	20.		20	
Magnesia 18.5	10.33	12.	*Grammatite.*	30	*Baikalite.*
Alumina ... 6.	....	....		...	
Ox. Iron....	.16	...		6	
Carb. Acid 26.	6.5	12. C. Lime }			
Loss 1.5	....	4.		....	

			Laugier.		
Silica 50	35.5	41.	28.4	
Lime 18	26.5	15.	30.6	
Magnesia 25	16.5	15.25	18.	*Grammatite.*
Water and Carb. Acid.... } 5	23.	23.	25.	
Loss 2	....	5.75	....	

Fibrous—asbestiform.

Species.   Hypersthene—*Labrador Hornblende.*

I.

	Klapr.	Theory ?	
Silica	........54.25	Bisil. Iron ....46.27 +	1
Magnesia	....14.	Bisil. Magn....35.38 +	1
Ox. Iron	....24.25	.... 1.0 +	
Alumina	.... 2.25	Silica .... 10.07 +	
Lime	.... 1.5	&c.	
Water	.... 1.		
Loss	.... 2.5		

II. *Prim.* Right rhomboidal prism, M. T. = 81° 48′.

III. Crystals 2—acicular—amorphous.

IV. S. G. = 3.3857.

Species.   Diallage—*Schiller spar.*

	Gmelin.	Drap.	Heyer.		Klap.			Vauq.
Silica	....43.7	41.	52.		60.			50.
Magnesia	11.3	29.	6.		27.5			6.
Ox. Iron	23.7	14.	27.5	Bronzite.	10.5	Smaragdite.		5.5
Alumina	17.9	3.	23.33		....			11.
Lime	....	1.	7.		....			13.
Water	....	10.			.5			4.5
Loss	....	2.			1.5			7.5
								Chrome.

II. *Prim.* Oblique irregular quadrangular prism, P. T. = 109° 18′.

III. Crystal 1—acicular—amorphous.

IV. S. G. = 3.

Species. Anthophyllite.

I.

		John.
Silica	62.66	56.01
Alumina	13.33	13.3
Lime	3.33	3.33
Magnesia	4.	14.
Ox. Iron	12.	6.
Ox. Mang	3.25	3.
Water	..	1.43
Loss	1.43	..

II. *Prim.* Right rhomboidal prism, M. T. $= 73^\circ$ 44'.

III. Crystal 1—acicular—amorphous.

IV. S. G. $= 3.2$.

Species. Epidote—*Thallite—Pistacite.*

I.

	Vauq.	Chen.	Cordier.
Silica	37.	45	33.5
Alumina	21.	28	15.
Lime	15.	15	14.5
Ox. Iron	24.	11	19.5
Ox. Mang.	1.5	..	12.
Loss	1.5	..	5.5

II. *Prim.* Right irregular four-sided prism, M. T. $= 114^\circ$ 37'.

III. Crystals 7 — granular — arenaceous — amorphous.

IV. S. G. $= 3.4529$.

vi. Zoisite; S.G. = 3.26 to 3.31 | Scorza; S.G. = 3.13

	Laugier.	Klap.	Klap.
Silica....	37.	45	43.
Alumina .	26.6	29	21.
Lime ..	20.	21	14.
Ox. Iron .	13.	3	16.5
Ox.Mang.	.6	..	.25
Water ..	5.8	..	..
Loss	1.	2	5.25

Species. Axinite—*Thumerstone.*

	Vauq.	Theory?		Klaproth.	
Silica	44	43.2	4	50 5	52.7
Lime	19	19.575	1	17.	9.4
Alumina	18	18.225	1	16.	25.6
Ox. Iron	14	&c.		9.5	9.6
Ox. Mang. ..	4			5.25	..
Potash				.25	..
Loss	1			..	2.7
Or,					
Bisil. Lime ..		41.175	1		
Bisil. Alum...		39.825	1		
&c........		....			

ii. *Prim.* Right irregular four-sided prism.

iii. Crystals 18—lamelliform.

iv. S. G. = 3.213.

Species. Lazulite—*Azurite*.

I.

	Tromsd.	Klap.
Silica	10.	46.
Alumina	66.	14.
Lime	2.	..
Carb. Lime	..	28.
Sulp. Lime	..	6.5
Magnesia	18.	..
Ox. Iron	2.5	3.
Water	..	2.

II. *Prim.* Rhomboidal dodecahedron.

III. Crystals 2—Amorphous *(Lapis Lazuli.)*

IV. S. G. = 2.767 to 2.945.

Species. Harmotone—*Cross stone*.

I.

	Keyer.	Klap.	Tassaërt.
Silica	44	49	47.5
Alumina	20	16	19.5
Barytes	24	18	16.
Water		15	13.5
Loss	12	2	3.5

II. *Prim.* Symmetrical octohedron, P. P'. $= 86^{\circ} 36'$.

III. Crystals 13.

IV. S. G. = 2.30 to 2.35.

Species. Meionite.

I.

Silica	58.7
Alumina	19.95
Potash	21.4
Lime	1.35
Ox. Iron	.4

Arfwedson.

II. *Prim*. Right symmetrical four-sided prism.
III. Crystals 3—In grains.
IV. S. G. = 2.612 to 3.1.

Species.   Nepheline—*Sommite.*

	Vauq.	Theory?		or		
Silica ..	46	44.71 +	3	Sil. Alum..40.0 +	1	
Alumina	49	50.28 +	2	Bisil. Al...54.9 +	1	
Lime ..	2	&c.				
Ox. Iron	1					
Loss ..	2					

II. *Prim*. Regular hexahedral prism.
III. Crystals 2—acicular—granular.
IV. S. G. = 3.2741.

Species.   Pinite.

I.	Klapr.	Drapier.
Silica........	29.5	46.
Alumina	63.75	42.
Ox. Iron	6.75	2.5
Loss		9.5

II. *Prim*. Regular hexahedral prism.
III. Crystals 3.
IV. S. G. = 2.92.

Species.   Iolite—*Dichroite—Cordierite.*

I. Silica........42.6 | 43.6
   Alumina ....34.4 | 37.6
   Lime ...... 1.7 | 3.1
   Magnesia .... 5.8 | 9.7 } Gmelin.
   Ox. Iron ....15. | 4.5
   Ox. Mang. .. 1.7 | ..
   Potash .... | 1.

II. *Prim.* Regular hexahedral prism.
III. Crystals 3—amorphous.
IV. S. G. = 2.56.

Species.   Laumonite—*Efflorescent zeolite.*

I. Silica........49.
   Alumina ....22.
   Water ......17.5 } Vogel.
   Lime ...... 9.
   Carb. Acid .. 2.5

II. *Prim.* Rectangular octohedron, P. M. = 117°
   2′;   P′. M′. = 98° 12′.
III. Crystals—acicular—aggregated.
IV. S. G. = 2.23 to 2.3.

Species.   Triclasite—*Fahlunite.*

I. Silica........46.79
   Alumina 26.73
   Water 13.5
   Magnesia .... 2.97 } Hisinger.
   Ox. Iron and } 5.44
   Ox. Mang.. }

II. *Prim.* Oblique rhomboidal prism, P. M. $= 99^\circ$
24′; M. T. $= 109^\circ$ 28′.

III. Crystal 1—amorphous.

IV. S. G. $= 2.6$.

Species. Prehnite—*Koupholite.*

I.

	Hassenf.	Theory?		Klap.	Theory?	
Silica ....50.	49.9+	4	42.87	41.94	3
Alumina....20.4	21.0+	1	21.5	23.6	1
Lime 23.3	22.6+	1	26.5	25.33	1
Magnesia .5	&c.		..	&c.	
Ox. Iron.... 4.9			3.		
Ox. Mang........			.25		
Water 9			..		
Loss			4.63		

II. *Prim.* Right rhomboidal prism, M. T. $= 102^\circ$
40′.

III. Crystals 5—fibrous—mamillary.

IV. S. G. $= 2.6097$ to 2.6969.

Species. Mesotype—*Needle Zeolite.*

	Smithson.	Geblen.		Pell.	Vauq.
Silica........49.	53.39	54.4	50	50.24
Alumina ....27.	19.62	19.7	20	29.3
Soda ....17.	14.69	15.09	....	....
Lime	1.75	1.61	8	9.46
Ox. Iron	..	..	....	....
Water 9.5	9.71	9.83	22	10.
Loss	..	..	....	1.

ii. *Prim.* Right rhomboidal prism, M.T. $= 93° 22'$.

iii. Crystals 6—fibrous and mamillary—pulverulent (*Mealy Zeolite.*)

iv. S.G. $= 2.083$.

vi. Natrolite; S.G. $= 2.2$.

Silica .... 48.	
Alumina 24.25	
Soda 16.5	Klaproth.
Ox. Iron 1.75	
Water .... 9.	

Species. Stilbite—*Foliated Zeolite.*

I.

	Gehlen.		Vauq.
Silica 55.07	55.61	52.
Alumina 16.58	16.68	17.5
Lime 7.58	8.17	9.
Soda (with some Pot.) } 1.5	1.53	....
Water 19.3	19.3	18.5
Loss	....	3.

ii. *Prim.* Right rectangular prism.

iii. Crystals 6—lamelliform—mamillary.

iv. S.G. $= 2.5$.

Species. Apophyllite—*Ichthyophthalmite—Fish-eye Stone.*

	Vauq.	Rose.		Reiman.
I. Silica 51	55.	52.	55.
Lime 28	25.	24.5	24.7
Potash 4	2.25	8.1	....
Alumina	....	....	2.3
Magnesia....	....	....	.5
Water 17	15.	15.	17.
Loss	2.75	.4	....

II. *Prim.* Right symmetrical four-sided prism.

III. Crystals 6.

IV. S. G. = 2.37 to 2.46.

Species. Chabasie—*Cubic Zeolite.*

I. Silica........ ....43.33
  Alumina 22.66
  Lime 3.34
  Soda (with some Pot.) .... 9.34   }Vauquelin.
  Water ....21.
  Loss 33

II. *Prim.* Obtuse rhomboid, P. P′. = 93° 48′.

III. Crystals 3.

IV. S. G. = 2.7176.

Species. Scapolite—*Paranthine.*

	Abild.	Berz.	Laugier.	Simon.	Ekeberg.
I. Silica......	48	61.5	43.	53.	46.
Alumina ...	30	25.75	33.	15.	28.25
Lime	14	3.	17.6	13.25	13.5
Magnesia....		.75	....	7.	....
Ox. Iron ...	1	1.5	.5	2.	.75
Ox. Mang...		1.5	.5	4.5	....
Soda........		...	1.5	3.5	5.25
Potash		....	.5	....	....
Water	2	5.	....	....	3.25
Loss	5	1.	1.4	1.75	3.5

II. *Prim.* Right symmetrical prism.

III. Crystals—acicular—amorphous.

IV. S. G. = 3.71.

APPENDIX 1.?

Wernerite.

I. Silica	51.5	50.25	40.	
Alumina ..	33.	30.	34.	
Lime	10.45	10.5	16.5	
Ox. Iron ..	3.5	3.	8.	John.
Ox. Mang..	1.45	2.45	1.5	
Alkali	....	2.	..	
Water	....	2.85	....	

II. *Prim.* Right symmetrical prism.

III. Crystal 1—amorphous.

IV. S. G. = 3.6063.

Appendix 2.?

Dipyre.

I. Silica......60 ⎫
   Alumina ...24 ⎪
   Lime......10 ⎬ Vauquelin,
   Alkali .... 2 ⎪
   Loss ...... 4 ⎭

II. *Prim.* Rectangular prism.

III. Crystal 1—acicular.

IV. S. G. = 2.6305.

Appendix 3.?

Gabronite.

	M'Kenz.	John.
I. Silica	71.17	54.
Alumina	13.6	24.
Potash	3.19	}17.25
Soda	....	
Lime	.4	....
Magnesia	...	1.5
Ox. Iron	1.4	1.25
Ox. Mang.	.4	...
Water	3.5	2.
Loss	6.64	....

III. Crystals ?—amorphous.

IV. S. G. = 3.

Species. Haüyne—*Latialite.*

	Vauq.	Gmelin.	
I. Silica ....30.	35.48	47.1	
Alumina ..15.	18.85	18.5	
Potash ..11.	15.46	6.4	
Lime .... 5.	2.66	5.4	
Sulphate of Lime } 20.5	21.73	....	
Ox. Iron.. 1.	1.16	13.7	
Water	1.2	....	
Sul. Acid..	....	1.2	
Loss 17.5	3.45	....	

II. *Prim.* Rhomboidal dodecahedron.

III. Crystal 1—granular *(Saphirine)*—massive.

IV. S. G. = 3.33.

Species. Analcime—*Cubic Zeolite—Sarcolite.*

	Vauquelin.		
I. Silica ....58.	50.	50.	
Alumina ..18.	20.	20.	
Soda ....10.	4.5	4.25	
Lime .... 2.	4.5	4.25	
Water .. 8.5	21.	20.	
Loss 3.5	....	1.5	

II. *Prim.* Cube.

III. Crystals 3—globular—amorphous.

IV. S. G. = 2. +

Species. Leucite—*White Garnet—Amphigene*
    —*Vesuvian.*

	Klaproth.			Vauq.
I. Silica.. ....53.75	54.23	54.5	56
Alumina 24.62	22.	23.5	20
Potash ....21.	22.	19.5	20
Lime	....	..	2
Loss .... .28	1.	2.5	1

II. *Prim.* Cube.

III. Crystal 1—rounded masses.

IV. S. G. = 2.45 to 2.49.

Species. Sodalite.

	Ekeberg.	Thoms.	Borkow.
I. Silica. ....36.	38.52	45.
Alumina ....32.	27.48	24.
Soda ....25.	23.5	27.
Lime	2.70	....
Ox. Iron .... .25	1.	.1
Water	2.1	....
Mur. Acid... .... 6.75	3.	....
Loss	1.7	3.9

II. *Prim.* Rhomboidal dodecahedron.

III. Crystals 2—amorphous.

IV. S.G. = 2. to 2.378.

APPENDIX ?

Lythrodes.

I. Silica.......... .. .44.62	
Alumina 37.56	
Soda .... 8.	John.
Lime 2.75	
Water .... 6.	
Ox. Iron .... 1.	

III. Disseminated—amorphous.
IV. S. G. = 2.5.

Species. Mica.

	Klaproth		Vauq.	Berz.
I. Silica...........47.	48.	42.5	50.	40
Alumina . ..20.	34.25	11.5	35.	46
Potash 14.5	8.75	10.	....	....
Lime.......	....	....	1.33	....
Magnesia	.5	9.	1.35	5
Ox.Iron........15.5	4.5	22.	7.	9
Ox. Mang.... 1.75	....	2.	....	...
Water	1.25	1.	....	...
Loss	....	....	5.33	....

II. *Prim.* Right rhomboidal prism, M.T. = 120°.
III. Crystals 5 — foliaceous *(Muscovy glass)* — lamelliform—pulverulent.
IV. S.G. = 2.53 to 2.93.

APPENDIX 1.

Chlorite.

	Lamp.	Vauquelin.		
I. Silica ....35.	56	26.	56
Alumina....18.	18	15.5	18
Potash	8	2.Mur. Pot.	8
Lime 29.9	3	....	3
Magnesia.	....	8.	....
Ox. Iron... 9.7	4	43.3	4
Water 2.7	6	4.	6
Loss 4.7	5	1.2	5

III. Crystals—earthy—amorphous.
IV. S. G. = 2.82.
VI. Chlorite slate.

Appendix 2.

Scaly Talc—*Nacrite*—*Talc granuleux.*

I. Silica........50.
  Alumina ....26.
  Potash ......17.5 }Vauquelin.
  Lime........ 1.5
  Ox. Iron .... 5.

III. Aggregated scales.

Appendix 3.

Lepidolite—*Lilalite.*

I.	Klap.		Vauq.	Troms.	John.
Silica 54.5	54.5	54.	52.	61.6
Alumina 38.25	38.25	20.	31.	20.61
Potash	4.	18.	7.	9.16
Lime...........	....	4.Fl.	8.5	1.6
Ox. Iron	.75	1.	.25	....
Ox. Mang....	....	3.	....	.5
Loss 6.5	....	....	1.25	1.86

III. Aggregated scales.
IV. S. G. = 2.81.

Species. Talc.

I.	Vauq.	Theory.		Klap.	Theory.	
Silica ......62.	62.8+	3	61.	64.5+	3
Magnesia ....27.	26.1+	1	30.5	26.9+	1
Potash	&c.		2.75	&c.	
Alumina 1.5			..		
Ox. Iron 3.5			2.5		
Water ...... 6.			.5		

ii. *Prim.* Right rhomboidal prism, M. T. = 120°.

iii. Crystal 1—scaly—pulverulent—amorphous.

iv. S. G. = 2.58 to 2.87.

vi. French Chalk.

Potstone *(Lapis ollaris.)*

APPENDIX 1.

Chlorite?

	foliated.	earthy.	
Silica	41.15	37.	
Magnesia	39.47	43.7	
Alumina	6.13	4.1	
Lime	1.5	6.2	Höpfner.
Ox. Iron	10.5	12.8	
Water	1.5	..	
Loss	.10	..	

APPENDIX 2.

Steatite—*Soapstone.*

	Vauq.	Klaproth.			Chenev.
Silica	36.	59.5	48.	45.	60.
Magnesia	37.	30.5	20.5	24.75	28.5
Alumina	..	..	14.	9.25	3.
Lime	2.	..	..	..	2.5
Ox. Iron	17.	2.5	1.	1.	2.25
Potash	..	..	..	.75	..
Water	4.	5.5	15.5	18.	..
Loss	6.4	2.	..	..	3.73

Agalmatolite.

Precious Serpentine.

	John.	
Silica	42.5	43.07
Magnesia	38.63	40.37
Alumina	1.	.25
Lime...	.25	.5
Ox. Iron	1.5	1.17
Ox. Mang....	.62	..
Water	15.2	..
Loss	.25⎱	12.45⎱
	Chrome⎰	Volatile⎰

Common Serpentine.

Verde Antique.

Species?   Asbestus.

I.	Chenev.	Berg.		Bergman.		
Silica	59.	64. ⎱		64.	63.5⎱	
Magnesia	25.	..		..	16.	
Lime	9.5	..	*Amianthus.*	..	12.8	*Hard Asbest.*
Carb. Mag...	..	17.2⎰		13.6	..	
Carb. Lime...	..	13.5		6.9	..	
Alumina	3.	2.7		3.3	1.1	
Ox. Iron	2.25	2.2		1.2	6.	
Loss	1.25	.. ⎰		Baryt.6.	....	

III. Fibrous.

IV. S. G. $= 0.9088$ to 2.9958.

vi. Rock-cork—Silica   56.2 to 62   per cent.
            C. Mag. 22. to 26   &c.

Rock-wood,   Silica 72. ,   Carb. Mag. 12.19
     per cent.

Mountain-leather.

ix. Ancient cloth.

Species.   Felspar.

I.	Vauq.	Klap.	Adularia—Moonstone	Vauq. 62.4	Vauq. 74.	Common.
Silica	64	68.		62.4	74.	
Alumina	20	15.		17.	14.5	
Potash	14	14.5		..	..	
Lime............	2	..		1.2	5.5	
Ox. Iron		.5		4.	..	
Water		..		15.4	..	
Loss		2.		..	6.	

ii. *Prim.* Oblique irregular four-sided prism, P.M.
= 90°; P. T. = 68° 20′; M. T. = 120′.

iii. Crystals 24—amorphous.

iv. S. G. = 2.56 to 2.74.

vi. Glassy felspar.
     Labrador felspar.
     Compact felspar—Saussurite; S. G. = 3.389.
     Decomposed felspar (*Kaolin*).

Appendix 1.?

Andalusite.

I.

	Vauq.	Guyton.
Silica	38	29.12
Alumina	52	51.07
Potash	8	....
Ox. Iron	2	7.83

III. Crystal—amorphous.

IV. S. G. = 3.16.

VI. Blue felspar; S. G. = 3.06.

Silica	14.	
Alumina	71.	
Potash	.25	
Lime	3.	} Klaproth.
Magnesia	5.	
Ox. Iron	.75	
Water	5.	
Chrome	.25	

Species. Petalite.

I.

	Clarke.	Holme.	Arfwcds.
Silica	80.	76.25	79.212
Alumina	15.	20.25	17.225
Lithium	..	..	5.761
Ox. Mang	2.5	2.25	....
Water	.75	.75	....
Loss	1.75	..	....

II. *Prim.* Right rhomboidal prism, M. T. = 137° 10′.

III. Amorphous.

IV. S. G. = 2.436 to 2.62.

Species. Spodumene—*Triphane.*

I.	Vauq.	Berz.	Vogel.	Arfwed.
Silica	64.4	67.5	63.5	66.4
Alumina	24.4	27.	23.5	25.3
Potash	5.	....	6.	....
Lithium	..	....	....	8.85
Lime	3.	.63	1.75	....
Ox. Iron	2.2	3.	2.5	1.45
Water	..	....	2.	....
Loss	1.	1.34	....	....
Volatile......	..	.53	....	....

II. *Prim.* Octohedron, P. M. $= 145° 42'$; P'. M'. $= 79° 50'$.

III. Crystal 1—amorphous.

IV. S. G. $= 3.192$ to 3.2.

MINERAL SPECIES NOT ANALYSED.

Species. Helvin.

II. *Prim.* Rhomboidal dodecahedron?
III. Crystal 1.
IV. S. G. $= 3.5$.

Species. Chiastolite—*Macle*—*Cross-stone*—
Hollow-spar.

II. *Prim.* Rectangular octohedron, P. P′. = 120°,
M. M′. = 98° 10′.

III. Crystals 4.

IV. S. G. = 2.944.

Species. Condrodite.

II. *Prim.* Oblique rectangular prism, P. M. =
112° 12′.

III. Crystal 1.—In grains.

IV. S. G. = 3.14.

Species. Spinellane.

II. *Prim.* Obtuse rhomboid, P. P′. = 117° 23′.

Species. Melilite.

Species. Pseudo-Sommite.

Species. Crichtonite.

II. *Prim.* Acute rhomboid, P. P′. = 18°.

Species. Humite.

Species. Ice Spar.

Species. Spinthere.

MINERAL MASSES OF HOMOGENEOUS TEXTURE,
WHICH DO NOT FORM TRUE SPECIES.

Obsidian—Marekanite—Pumice.
Pearl-stone.
Pitch-stone.
Clink-stone.
Basalt.
Common Jade—Axe-stone.
Green Earth.
Slates.
Clays.

INDEX.

I

114 INDEX.

Printed in the United States
By Bookmasters